Divided Not Conquered

Divided Not Conquered

How Rebels Fracture and Splinters Behave

EVAN PERKOSKI

OXFORD
UNIVERSITY PRESS

Oxford University Press is a department of the University of Oxford.
It furthers the University's objective of excellence in research, scholarship,
and education by publishing worldwide. Oxford is a registered trade mark of
Oxford University Press in the UK and certain other countries.

Published in the United States of America by Oxford University Press
198 Madison Avenue, New York, NY 10016, United States of America.

© Oxford University Press 2022

Library of Congress Cataloging-in-Publication Data

Names: Perkoski, Evan, author.
Title: Divided, not conquered : how rebels fracture and splinters behave /
Evan Perkoski.
Description: New York, NY : Oxford University Press, [2022] |
Includes index.
Identifiers: LCCN 2022018897 (print) | LCCN 2022018898 (ebook) |
ISBN 9780197627075 (paperback) | ISBN 9780197627068 (hardback) |
ISBN 9780197627099 (epub)
Subjects: LCSH: Terrorist organizations. | Terrorists—Psychology. |
Social groups. | Group values (Sociology)
Classification: LCC HV6431 .P463 2022 (print) | LCC HV6431 (ebook) |
DDC 363.325—dc23/eng/20220525
LC record available at https://lccn.loc.gov/2022018897
LC ebook record available at https://lccn.loc.gov/2022018898

1 3 5 7 9 8 6 4 2

Marquis, Canada

For my parents

Contents

Acknowledgments

This book began as my doctoral dissertation at the University of Pennsylvania. I was fortunate to have Mike Horowitz, Jessica Stanton, and Avery Goldstein guiding my research. They made my graduate experience enjoyable and I model myself on the precedent they established. I am especially indebted to Mike who saw my potential as an eager undergraduate and who has since advised me through every facet of my career.

This book has even deeper roots that stretch back to my undergraduate education at Wesleyan University where I received a solid foundation for my graduate coursework. I am grateful to all of the faculty who taught me, but especially Erica Chenoweth whose course on the Politics of Terrorism convinced me that political science was more interesting than law. Erica has been a consummate adviser, role model, and friend.

Numerous institutions supported me writing this book. I finished my dissertation while a fellow at the Harvard Kennedy School's Belfer Center for Science and International Affairs. Steve Walt and Steve Miller provided valuable feedback and were exceptional role models. I am also grateful for the support of my Belfer colleagues including Mark Bell, Kelly Greenhill, Morgan Kaplan, Peter Krause, Julia MacDonald, Rupal Mehta, Mike Poznansky, Gaëlle Rivard-Piché, Rachel Whitlark, Alec Worsnop, and others. I then refined this project while a fellow at the Sié Chéou-Kang Center for International Security and Diplomacy at the Josef Korbel School for International Studies at the University of Denver. This was a professionally and personally rewarding experience, made so by the likes of Debbie Avant, Marie Berry, Erica Chenoweth, Cullen Hendrix, and Ollie Kaplan. This project culminated when I was an Assistant Professor at the University of Connecticut. UConn provided the support and the flexibility to cross the finish line. Finally, I am also thankful for support from the Smith Richardson Foundation, the Horowitz Foundation for Social Science, the Browne Center for International Politics, and Penn's School of Arts and Sciences.

List of Tables

List of Figures

1

Introduction

Terrorists, insurgents, and rebels frequently experience internal disagreements that tear their organizations apart. Dissatisfied members, and usually those on the losing side of an argument, will commonly break away to form independent groups of their own—splinters, as they are often called. In this way, armed groups are no different from other human organizations that "form, transform, breakup, and reform at a speed that has no parallels in the animal world."[1] But splinters that break away from armed groups exhibit wildly different patterns of behavior. Some thrive and endure for decades; others quickly fade away. Some also radicalize and launch violent, destructive attacks; others still moderate and pose little threat of their own.

To illustrate, consider the following two cases. First, the roots of the Irish Republican Army (IRA) can be traced to 1913 when a group of militant nationalists formed the Irish Volunteers. Despite their fame and at times widespread support, the IRA was perpetually beset by internal disagreements and organizational fractures. These disagreements spawned myriad splinters from the ranks of the IRA's well-trained cadre. Far from being inconsequential, these fractures fundamentally shaped the conflict's trajectory and produced some of the deadliest and most durable organizations of that era, including the likes of the Official IRA, the Real IRA, the Provisional IRA, and the Continuity IRA, among many others.[2]

Second, splits were also frequent among Palestinian militant groups. One noteworthy rupture divided the highly successful Popular Front for the Liberation of Palestine (PLFP), whose leader, Waddi Haddad, was responsible for important tactical and strategic innovations.[3] Specifically, Haddad oversaw the first hijacking of an Israeli airliner in 1968, and in 1970 he directed the simultaneous hijacking of four separate planes.[4] Despite his success, the PLFP ruptured in 1978 and three splinters emerged: PLFP-Special Command, the May 15th Organization, and the Lebanese Armed Revolutionary Factions. Together, these groups only ever accounted for 17 attacks and three fatalities, and they were all inactive within five years.[5] Far

Divided Not Conquered: How Rebels Fracture and Splinters Behave. Evan Perkoski, Oxford University Press.
© Oxford University Press 2022. DOI: 10.1093/oso/9780197627075.003.0001

from being significant, as many were in Northern Ireland, these breakaway groups were surprisingly ephemeral.

The cases of the Irish Republican Army and the Popular Front for the Liberation of Palestine demonstrate the very different outcomes that are possible when armed groups break apart. They also bring to mind two of the popularized yet competing conceptions of organizational splintering. First, there is a widespread notion that splinter groups will be highly radicalized and dangerous. This was certainly the case with some IRA factions, including both the Real IRA and the Continuity IRA, that embraced increasingly violent tactics and strategies. Accordingly, models of group fragmentation often assume that it is the most hardline terrorists and insurgents who break away, rejecting compromise and leaving their more moderate peers behind. This idea has even made its way into popular culture. For instance, the eight novels comprising the *Tom Clancy: Splinter Cell* series depict rogue CIA agents seeking to undermine the global order. News outlets have done their part to reinforce this association as well, with headlines like "Hard Line Splinter Group . . . Emerges from Pakistani Taliban,"[6] and "Splinter Terrorist Cells May Target Election."[7]

Yet, there is another conception of splinter groups that is more closely related to the experience of the PFLP, a case where several groups emerged and then quickly disappeared. Here, the association is more with "fragmentation" than with splintering. While the latter often conjures images of radicalization and extremism, fragmentation is often linked to weakness, decline, and a loss of control—even though they describe the same underlying events. To this point, the very idea of an organization breaking apart, whether it is a militant group, a political party, or a business firm, suggests a troubling lack of control and consensus that reflects weakness and turmoil. Counterinsurgent forces' attempts to divide militant organizations is based upon this very notion—that splintering is either a cause or a product of organizational weakness and that breaking a group apart can help to defeat it.

These competing conceptions are both flawed. Not every splinter group will radicalize and alter the conflict landscape, nor is a splinter's emergence necessarily indicative of organizational weakness and counterinsurgent success. This dynamic frames the question at the center of this book: what explains variation among splinter organizations? Specifically, why do some splinter groups survive and even thrive while others die out relatively quickly? And why do some groups radicalize and adopt tactics and strategies that are increasingly violent? From a policy perspective, these questions are critical to designing effective counterterrorism and counterinsurgency strategies. While armed groups often break apart on their own, owing to natural organizational dynamics, states commonly seek to accelerate this process through acts of repression and even conciliation. However, states will want to avoid fracturing militants in ways that produce durable and radical offshoots, like some of those from the Irish Republican Army, and instead cultivate fissures that promote weakness and disarray, like that which inflicted the PFLP. From a research perspective, there are few explanations for why splinter groups exhibit such distinct trajectories, but there are many more for why groups break apart. There is also a great deal of research on how armed groups behave, why they radicalize, and why they survive or fail. Research at the intersection of these topics is entirely missing, however, leaving an important gap in our understanding of armed groups and subnational conflict.

But filling this gap is important. For one, this process is incredibly common. Researchers agree that approximately one-third of all armed groups form by breaking away from preexisting organizations, so this topic has wide-reaching implications.[8] In addition, the groups that emerge from organizational fractures are themselves quite significant. They form with experienced members, strategic know-how, and will often compete against their parents for control, respect, and recruits. The emergence of a new splinter group can therefore have important ramifications for human security and conflict dynamics.

In this book, I show how intraorganizational politics are essential to understanding the breakdown of armed groups. Specifically, to understand how splinter groups behave and survive, we must first understand their motivations for breaking away. This is because the disagreements precipitating organizational splits influence which fighters stay behind with the parent group and which leave to join the splinter. Disputes over strategy, for instance, typically attract the most hardline members away from the parent organization and are therefore most likely to create splinters that radicalize, attack rivals, and undermine peace. Disagreements over ideology and leadership, meanwhile, are less likely to produce new organizations that ramp up violence and escalate their operational profiles. The reasons motivating a group to break apart also affect the resilience of splinters. Here, internal consensus is key. Splinter groups that form around a single issue tend to attract a more like-minded base of recruits who all want to reform their organization in similar ways. This internal homogeneity and preference alignment facilitate coordination, cooperation, and control, all of which make a splinter group more likely to survive its tenuous first days. Taken together, this book reveals how the disagreements within armed groups shape the membership composition of breakaway splinters, and how their membership in turn shapes their behavior and survivability.

I prove my argument using both both qualitative and quantitative evidence. Utilizing a new data set, I statistically analyze three hundred armed organizations to explain patterns of behavior and survival. The results show that the disagreements underlying a splinter group's formation shape its trajectory for years to come. I also demonstrate the explanatory capacity of my argument through in-depth case studies of the Basque terrorist group, Euzkadi Ta Askatasuna (ETA), the Irish Republican movement, and the Islamic State. I draw upon original research from Belfast, London, and Dublin to study the IRA, and from Madrid for ETA. All three cases show how splinter groups are products of the internal political forces that led them to emerge.

Accordingly, this book understands militant groups through a rationalist, organizational lens. Rather than viewing armed groups and their members as crazed fanatics bent on destruction, I show how a group's behavior is the product of predictable forces. In this sense, armed groups are no different from other groups in society. While it is no surprise that we can understand aspects of a business—be it Facebook or Apple or Ford—by examining its internal structure, its leadership, its market, and its competitors, the same is true of groups of men and women seeking to affect political change through the use of force. Their internal structures (e.g., hierarchy versus decentralization), leadership (e.g., factionalized versus unified), markets (population of supporters), and competitors (other armed groups) all shape their tactics, strategies, alliances, and other behaviors. This particular approach to studying armed groups has already revealed much about their behavior, and there is even more that it can help us to understand.

The contributions of this book are numerous. The insights improve our capacity to anticipate the trajectory of a consequential, sizable set of armed groups. My research speaks most directly to those that form by breaking away from preexisting armed organizations, shedding light on which are likely to destabilize peace process, to endure and threaten security for years to come, and which are likely to collapse on their own. In this way, the findings offer useful insights to policymakers tasked with countering threats from armed organizations. I show how certain types of organizational fractures have higher odds of producing more cohesive, radicalized groups.

These fractures should be avoided, while interventions that might lead to less stable and more tenuous splinter groups should be pursued. This information can also be used in negotiations with armed groups. I show how we can leverage insurgents' internal disagreements to logically infer who is departing from the parent organization. In cases where hardliners are splitting off, parent groups are more likely to negotiate and perhaps more able to enforce subsequent agreements. Lastly, this book underscores that the disputes within armed groups matter. For decades, research on organizational breakdown has not been studied as closely as organizational formation. This book shows what we stand to gain by taking these disputes and these fractures seriously.

The Fragmentation of Subnational Conflict

Conflict is a defining feature of human history that is perpetually evolving. Perhaps most notably, conflicts of today are no longer dominated by nation-states as they were for centuries. While fears of the potential devastation wrought by great-power wars are well placed, it is more common in the twenty-first century for states to use their armed forces against adversaries with no formal states of their own—like Al-Qaeda, the Islamic State, Boko Haram, and many others.

Figure 1.1 uses data from the Correlates of War project to plot the frequency of states engaged in wars against two types of opponents: other states (dotted line) and nonstate actors (solid line). For wars to be included in either of these categories they must produce at least 1,000 combat deaths in a single year and involve at least one state as a belligerent.[9] As this demonstrates, the number of states actively fighting other states peaked in 1941 during the early phases of the Second World War, when 22 different states were engaged in sustained combat.[a] Other peaks were reached, not surprisingly around 1970 when anticolonial conflicts and Cold War proxy battles were frequent. The number of states fighting wars against nonstate actors, on the other hand, peaked more recently. In 1992, a record high of 32 states were fighting wars again stateless opponents,[b] and this trend has been steadily rising since about

---

[a] This included France, Thailand, China, Japan, Australia, Bulgaria, Canada, Ethiopia, Finland, Germany, Greece, Hungary, New Zealand, Romania, South Africa, the Soviet Union, the United Kingdom, the United States, and Yugoslavia. Some of these states, like France and China, were engaged in several wars concurrently.

[b] This includes Algeria, Angola, Croatia, Bosnia, Papua New Guinea, Yugoslavia/Serbia, Moldova, El Salvador, Sierra Leone, Sri Lanka, Georgia, India, Mozambique, Tanzania, Zimbabwe, Azerbaijan,

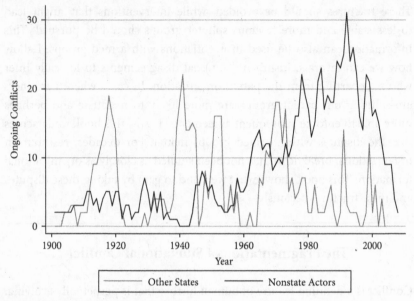

Fig. 1.1 The targets of state violence

1950. In fact, wars between states and nonstate actors have been the modal form of conflict since the inflection point in 1974.

While the broader trend in warfare has been a shift away from interstate disputes toward those between states and nonstate actors—what we can call intrastate or subnational conflicts[c]—these wars have also been changing in important ways.

First, the number of battle-related deaths from subnational conflicts has increased dramatically since the mid-2000s. This is evident in Figure 1.2. The gray area represents the distance between the lowest estimates and highest estimates for battle-related deaths from subnational conflicts in a given year, and the red line represents the best guess. The data imply there was a relative lull in deaths following the end of the Cold War, reaching a minimum of around 28,000 in 1996. This is not all that surprising, since Figure 1.1 also shows a decrease in state-nonstate conflicts around this time. This number spiked in 1999 with 102,000 deaths in a single year before dropping again,

the Philippines, Afghanistan, Liberia, Nigeria, Pakistan, Australia, Italy, Somalia, Nigeria, France, Sudan, Turkey, Peru, Tajikistan, and the United States.

[c] Of course, not all of these conflicts are truly contained within one state. Some may even involve one state attacking a nonstate actor in a foreign state. I use the terms intrastate and subnational to describe all of these interactions.

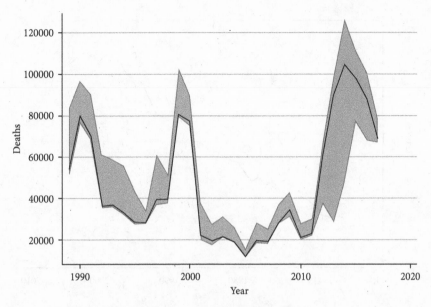

Fig. 1.2 Battle-related deaths from subnational conflicts

and then increased to its highest post–Cold War level in 2014 with over 126,000 fatalities.[d]

Second, the number of armed groups involved in conflicts around the globe has steadily increased as well, and usually faster than the number of conflicts. Data suggest that individual conflicts are now characterized by a multitude of independent armed groups vying against states, and in many cases against one another as well. For instance, among self-determination disputes in particular, Cunningham, Bakke, and Seymour Find that oppositions are unitary—represented by one armed group—in only 30 percent of cases between 1960 and 2005.[10] And, the number of factions in a single conflict has also steadily increased over time, which is displayed in Figure 1.3. In 1960, it was relatively common for self-determination movements to be led by one or two groups. In 2005, that number has risen to between four and five groups (4.23 to be exact). In other words, the average number of groups involved in secessionist wars has nearly doubled in the last 45 years. As one

---

[d] Conflicts in 2014 that contribute to this statistic include those in Syria, Iraq, Afghanistan, Nigeria, Pakistan, Ukraine, Israel, South Sudan, and Yemen. Although battle-related deaths have decreased since 2014, in 2017 subnational conflicts still managed to generate a staggering number of fatalities–nearly 80,000 in one single year. These conflicts exact a massive toll on human lives.

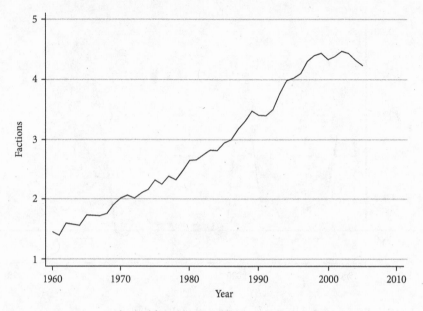

Fig. 1.3 Average number of factions in self-determination disputes

might imagine, researchers find that the proliferation of armed groups has a profoud impact on conflict dynamics.

This trend in the proliferation of armed groups is not limited to self-determination disputes, either. The same holds true for conflicts involving terrorist organizations. Though it is more difficult to link transnational terrorists to particular conflicts, as we can do with secessionist campaigns, we can instead look at their proliferation[e] within states more generally. Figure 1.4 plots the average number of active terrorist organizations per country in a given year.[f] In 1970, countries experiencing terrorist violence faced, on average, about three different organizations. Some of the most recent data suggests this number is now over five.

Taken together, the conflicts of today are dramatically different from those in previous decades. Whether it is civil wars, insurgencies, or terrorist campaigns, armed groups are proliferating and states are facing deadlier and more complex threats. There is an urgent need to better understand the processes underlying this important transformation.

[e] Data comes from the Global Terrorism Database. Although many of these groups are not what researches would label as "terrorists." Some have fought relatively complex insurgencies, like the Islamic State, while others are indeed typical terrorist organizations according to popular notions.

[f] For states that experience at least one terrorist attack in a given year.

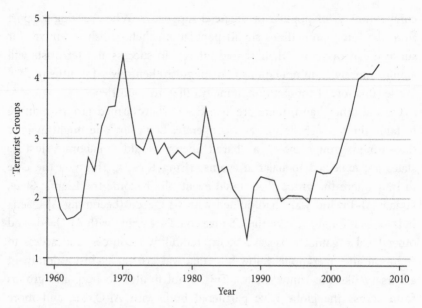

Fig. 1.4 Average number of terrorist groups per country over time

The Consequences of Fragmentation

Why does it matter that there are more armed groups fighting in more subnational conflicts? It matters because fragmented conflicts are fundamentally different. Specifically, the proliferation of armed groups alters two basic relationships that policymakers and researchers often take for granted: first, relationships between militant organizations (the different armed groups operating on the ground) and second, relationships between armed groups and the states they are fighting.[11]

Relationships between the armed groups operating in a given conflict have been under-theorized for a long time.[g] When it comes to how they will interact, researchers commonly default to an essentially realist perspective, assuming that groups will shun their shared interests to compete (directly or indirectly) with one another over limited gains. This was the case in Palestine with groups like Hamas, Fatah, and Palestinian Islamic Jihad. Studying how these organizations behave, Bloom writes that terrorist groups indirectly compete or try to "outbid" one another "when they are in

[g] This is especially true when compared to other dyads in a conflict scenario, such as state allies engaged in a multiparty interstate war.

competition...for popular or financial support."[12] When seeking support from the same (sometimes small) population, their selfish incentives for survival can overcome their shared interest in success and terrorists will employ more violent operations to distinguish themselves from the pack.[13] We see this sort of competition in many different countries.[14]

On the other hand, there are also cases where armed groups manage to tame their selfish desires and cooperate. Logically, one might assume that armed groups facing a shared enemy would collaborate, just as states are expected to balance against rising threats. This was the case in Iraq where the group that would eventually become the Islamic State, Jama'at Al-Tawhid wal-Jihad, joined with Al-Qaeda (becoming Al-Qaeda in Iraq) as it fought against the US invasion. Partnering with Al-Qaeda and other local organizations gave them credibility, resources, and access to technical and operational know-how that was critical against a dominant adversary like the United States. This is not limited to Iraq, and groups from across the globe have partnered both with Al-Qaeda and more recently with the Islamic State to reap similar benefits.[15] Accordingly, new analyses reveal that alliances between armed groups are increasingly common.[16]

While it is typical for researchers to take one of these two approaches, believing that armed groups typically compete or typically cooperate, both are overly simplistic. Armed groups are not simply interchangeable actors with mutable interests and characteristics that universally behave one way or another. While basic needs and the desire to survive mean that we can assume some fundamental similarities, we must also recognize that these are complex organizations that will defy attempts to stereotype their behavior. For instance, most research overlooks the basic familial links that exist between specific pairs of militant organizations and that introduce unique dynamics into a conflict. In some cases these links—like those between a parent group and a splinter—might exacerbate intergroup tensions and result in even greater competition. Indeed, spurned parents might even seek to eradicate their progeny and to directly fight for survival, control, and local superiority. This was the case in Ethiopia when the Eritrean Liberation Front launched an assault on the Eritrean People's Liberation Front, an amalgamation formed when three of its splinters merged.[17] In other cases, the enduring familial links between organizations might eventually foster cooperation in the face of a shared enemy or a threatening newcomer. Boko Haram, for instance, experienced a rupture that spawned the breakaway

Ansaru organization, but the two eventually reconciled and now operate jointly.[18] As these examples demonstrate, the dynamics among a pair of organizations will likely be different if one formed by breaking away from the other, and this would be overlooked if one were to solely focus on the number of active organizations. In effect, these underlying links challenge our assumptions about organizational behavior in multiparty conflicts.[19]

The second dynamic that is complicated by fragmentation is the relationship between armed groups and the state. States facing a fragmented opposition can no longer concentrate their attention on a single opponent. Resources must be apportioned and strategies re-imagined to contend with a multitude of competing organizations that have different internal structures, objectives, recruitment patterns, and levels of support. Designing an effective, comprehensive counterinsurgency strategy becomes significantly more difficult under these circumstances. The United States found itself in this situation during the mid-2000s when forced to contend with armed groups representing Shiite Arabs, Sunni Arabs, and Iraqi Kurds. As one RAND report notes, "Were insurgency the only challenge, U.S. and Iraqi government forces might at least contain the violence, but the multiple challenges of separatists, insurgents, extremists, militias, and criminals threaten to destroy the country at any moment."[20]

In line with the US experience in Iraq, research finds that fragmentation affects a host of conflict outcomes, including the likelihood and the efficacy of negotiations with armed groups. To this end, both Nilsson[21] and Driscoll[22] find that governments in fragmented conflicts have incentives to win over armed groups with conciliatory offers. States see an opportunity to decrease the number of active fronts by essentially buying off smaller groups to focus on larger, more threatening adversaries. Cunningham finds nearly the same dynamic at work, with evidence suggesting that states are more likely to provide concessions to divided rather than unified movements.[23] Here, the logic is that concessions are part of the bargaining process, used strategically "without attempting to settle underlying disputes."

While fragmentation therefore affects the likelihood and logic of conciliation, it also affects the odds of conflict termination. Greater numbers of groups complicate efforts to attain comprehensive peace deals that can actually settle a conflict. With more organizations, finding an optimal outcome that is agreeable to all becomes even more challenging. A divided opposition might also make it harder to find a viable negotiation partner. This happened in Syria where "The United States and its donor partners viewed the Syrian opposition's fragmentation as a major weakness in its potential to emerge as a viable counterstate."[24] Of course, more armed groups also means more potential spoilers. In these cases, spoilers are "leaders and parties who believe that peace emerging from negotiations threatens their power, worldview, and interests, and use violence to undermine attempts to achieve it."[25] Spoilers affect conflict outcomes when they maintain the ability to unilaterally continue a conflict, thereby disrupting and possibly ending ongoing negotiations.[26] Terrorist and other militant groups sometimes splinter when they are presented with a negotiation since the offer divides hardliners and moderates within an organization; moderates will favor the terms of the agreement, while hardliners prefer to keep fighting.[27] This can make conflict termination nearly impossible.

Taken together, fragmented conflicts generate what Bakke, Cunningham, and Seymour call a "dual contest"—the first between competing militant groups and the second between militants and the state.[28] But more than that, fragmentation fundamentally affects the dyadic relationships between pairs of armed groups, defying our attempts to generalize their behavior. Understanding these dynamics is critical in an era when subnational conflicts have become increasingly fragmented and armed groups are constantly emerging and breaking apart.

How Fragmentation Occurs and Armed Groups Emerge

With conflicts growing increasingly fragmented, where are these armed groups coming from? As I just showed, there can be distinctive conflict dynamics when a splinter group and its parent operate in the same environment, but this is only one of several means by which new armed groups come to exist. Table 1.1 presents a typology of how armed groups might emerge.[h]

First, new groups form through processes of organizational proliferation. This is where they emerge absent any direct lineage in another armed organization. Instead, they emerge from the ground up or from some preexisting group that does not currently engage in violence like a political party, trade union, or other social organization.[29] The key characteristic here is the lack of a violent antecedent. While a group's members might have experience in other armed groups and from previous conflicts, there is no direct

---

[h] In this typology I focus on how additional armed groups may enter a conflict, so it excludes merging which would inherently decrease the number of organizations. Little research exists on why organizations merge and how these new organizations fare.

Table 1.1 A typology of how armed groups emerge

Proliferation	New, independent armed groups that form without any preceding violent organization
Expansion	Existing armed group spreads to new location but maintains control
Decentralization	Existing armed group spreads to new location with significant local autonomy
Specialization	Intentionally created semi-autonomous subgroups (e.g., militant wings, death squads)
Splintering	Segment of preexisting armed group breaks away to become independent

organizational lineage tying them to an active or inactive violent entity. This was the case when the Weather Underground formed in the United States. While many of its members were previously part of the nonviolent Students for a Democratic Society, this was their first time participating in a violent organization. Unsurprisingly, they experienced many initial setbacks and faced a steep learning curve. While experimenting with home-made bombs in 1970, for instance, an accidental explosion killed three and leveled an apartment.[30]

Second, armed groups can emerge through organizational expansion. This is when an established group directly inserts itself into a new conflict, largely preserving its chain of command while simply expanding its base of operations. This scenario is perhaps most likely when an armed group operates in a country that borders another with an intrastate conflict owing to the relative ease with which members can travel back and forth. Organizational expansion becomes even easier when that land border contains mountains, jungles, or other obstacles that make it harder for states to monitor.[31] One of the most noteworthy cases in this regard is the Islamic State: after forming in Iraq and then joining Al-Qaeda, the group broke away and later expanded into ungoverned territory in Syria. Facing little resistance in the wake of the Syrian civil war, there were few deterrents to their regional spread.

Third, new armed groups can enter a conflict through processes of organizational decentralization. This is where an existing armed group purposefully establishes another violent organization in a separate location with significant autonomy. Perhaps they commandeer a local affiliate, pledging funds and sending members to seed the new entity. Nonetheless, the autonomy of this new organization is what differentiates decentralization from expansion. Aptly, some researchers call this strategy "franchising."[32] Mendelsohn, for instance, describes franchised insurgent groups as containing "a central managerial level and, below it, branches, each responsible for a particular geographical area."[33] Among others, Al-Qaeda embraced this approach beginning in the late 2000s.[34] Testifying at a congressional hearing in 2012 after the death of Osama Bin Laden, Director of National Intelligence James Clapper noted that:

> sustained pressure from the United States and its allies will probably reduce Al-Qaeda's remaining leadership in Pakistan to "largely symbolic impor- tance" over the next two to three years as the terrorist organization frag- ments into more regionally focused groups and homegrown extremists.[35]

Why might armed groups choose to decentralize their operations and not expand across territory and maintain control? The main reason has to do with managing risk. Decentralized networks are safer, and they are harder to destabilize and decapitate than hierarchical networks that preserve top-down control.[36] In addition, maintaining hierarchy across state lines is simply difficult owing to the inherent challenges of communication, supply, and troop movement. Decentralization is therefore a smart choice for expanding armed groups. As Woldemarian points out, however, "To the outside observer, this reduction in coordination across rebel units may be interpreted as the rebel organization having fragmented, but to the parties involved, this is a tactical maneuver that has little to do with any underlying factional dispute."[37] That is, delegating autonomy may appear like an organizational rupture, when in reality decentralization can be both strategic and intentional while making the group even more resilient.

Fourth, militant groups sometimes face incentives to create specialized semi-autonomous entities that remain broadly under their control but that conduct especially radical operations. Although there is little research into this precise topic,[38] an armed group might create these types of organizations for two reasons. First, when a majority of an organization's supporters do not condone more radical operations (like suicide bombing campaigns), an armed wing provides a level of deniability and distance while also achieving the desired tactical capabilities. Second, groups might form armed wings to create avenues for more radical supporters to join their organizations over another. This strategy allows a group to maintain control over these individ- uals and to increase their own numbers while simultaneously undercutting their rivals. Without an armed wing, a group might lose out to rivals who cater to this population. Overall, these types of armed groups are distinct due to their deliberate creation to fulfill an organizational imperative.

An example of one of these semi-autonomous organizations comes from Palestine, specifically Fatah and its militant wings of Tanzim and the Al-Aqsa Martyrs Brigade. Fatah was founded in 1959 as a political movement seeking Palestinian independence, becoming an official Palestinian political party in 1965. Fatah has been careful to balance between militant organization and political party since swaying too far to one side will isolate a significant portion of the Palestinian community, compromise its credibility, or incite Israeli aggression. It has created distinct armed wings designed to ameliorate this strategic dilemma. Moghadem notes that "Fatah's formation of the Tanzim must be seen in the context of its attempt to channel and focus the passions of many Palestinians in the West Bank and Gaza away from Islamist groups."[39] The creation of Tanzim therefore enables Fatah to win recruits it would otherwise be unable to attract. The Al-Aqsa Martyrs Brigade was formed for similar reasons, and both demonstrate why the formation of militant wings cannot be viewed as signals of organizational breakdown or collapse.[40]

Fifth, and finally, is organizational splintering—the type of fragmentation motivating this study.[41] This is when a subset of members from an established armed organization break away to create a new group that is entirely independent. What distinguishes this phenomenon from militant wings and other forms of organizational fragmentation is autonomy: splinters are independent and no longer under their parent's jurisdiction or control. Militant wings like Tanzim and Al-Aqsa might receive some operational autonomy, but their parent maintains executive constraint on their activities while often providing funding, weapons, and recruits. A splinter organization, on the other hand, fends for itself.[i]

Even with these criteria, it can sometimes still be difficult to discern which is the splinter group and which is the parent, especially in low-information environments. Splinter groups may even obscure this intentionally to claim their organization's name for themselves, bringing with it familiarity and notoriety. In these cases, the breakaway segment will claim they are the "true" group while leaving behind those who have deviated from the rightful path. For instance, with the prospect of a negotiated settlement to the conflict in Northern Ireland in the 1990s, dissenters broke away claiming they were

---

[i] Another defining characteristic of splinters is that they are formed by a minority of group members. In cases where a majority of an organization departs, it raises the question of which of the two groups should be called the splinter, and those situations are more akin to an organizational evolution rather than splintering.

"the real IRA." They subsequently became known as the Real IRA or simply "the Reals." However, in cases where a splinter fundamentally disagrees with the parent group and wants to alter strategy or ideology—perhaps in the wake of an exceptionally indiscriminate or unprovoked attack—it may seek to distance itself. In such cases the splinter might benefit from a blank slate and by not carrying over the damaged reputation of its predecessor.

What We Think We Know about Organizational Splintering: Five Myths and Misconceptions

While splintering frequently occurs among armed organizations, we lack systematic research about it. In its place, a number of myths and misconceptions have developed to help make sense of these events and to understand their implications. In this section, I evaluate some of the most pervasive myths and I consider whether they in fact hold any validity. While some hold truth, others are simply unfounded.

Myth 1: It's Always Hardliners

The first and perhaps most widely held myth about organizational splintering is that hardliners are always to blame. In other words, many believe that it is an armed group's most radical members that rupture organizational unity and set off on their own. This assumption is held by researchers,[42] members of the public, the press, and even policymakers, and it perpetuates the idea of splinter organizations as inherently violent, radical, and anti-compromise.

This prevalent misconception owes its popularity to a few facts. The first is that disputes over strategy and the use of violence are among the most common grievances causing armed groups to break apart. However, they are not the only reason: in data that I describe more fully in the coming pages, I find that strategic disputes are involved in nearly 60 percent of organizational fractures, leadership disputes in 35 percent, and ideological disputes in 30.[j] It is also important to recognize that disagreements within armed groups over strategic choices are usually, but not always, about using more violence. A minority occur when dissatisfied members seek to moderate their use of force. For example, Ansaru broke away from the well-established

---

[j] This sums to over 100 percent since fractures can have multiple causes.

Boko Haram in 2012. Announcing their formation, the group released flyers around the city of Kano declaring itself a " 'humane' alternative to Boko Haram that would only target the Nigerian government and Christians in 'self-defense.' "[43] Some have cited Boko Haram's suicide operations against Muslim civilians as a key reason for this split.[44] As this demonstrates, some organizational ruptures could be instigated by moderates who precipitate tactical moderation.

The second reason for this myth has to do with media reporting. Research underscores that media coverage of violence, terrorism, and insurgency is far from unbiased. For instance, attacks committed by Muslims received 357 percent more coverage than equivalent attacks committed by non-Muslims.[45] Greater numbers of civilian casualties and particular targets (e.g., police) also shape rates of coverage. Other research finds similarly bias regarding the methods of violence as well,[46] with suicide attacks garnering significantly more press. As a result, certain types of splinter organizations are more likely to capture the media's interest: specifically, those that radicalize and employ increasingly heinous methods of violence. When these particular breakaway groups dominate the headlines, crowding out groups seeking to compromise with the government or groups breaking away for perceived organizational mismanagement, people will come to believe this behavior and these motivations are the norm.

Accordingly, this myth misrepresents the complex reality of how and why armed groups splinter. While strategic disagreements commonly underlie organizational fractures, they also split over other issues and only some of these portend escalation and radicalization.

Myth 2: Splinters Are Irrational and Unhinged

A second myth is that organizational splintering is a chaotic, haphazard event driven by unbalanced terrorists and insurgents whose organizations are failing. Some of this perception comes from the fighters themselves. In the wake of an organizational split, it often benefits one group—usually the parent—to label the breakaway faction as unhinged or murderous. This can be a useful strategy to deter support and undercut the foundation of the breakaway organization. For instance, this was Al-Qaeda's strategy in the wake of its split with the Islamic State. A 2013 audio message from Ayman Al-Zawahiri (then leader of Al-Qaeda) stated that "ISIS was struck with madness," and he later said that the group "[exceeds] the limits of extremism."[47]

Governments are similarly prone to playing up the irrationality of their adversaries. In the wake of major setbacks experienced by the Islamic State, for instance, President Trump stated that:

> You never say you won, because you have a couple of whack jobs go and blow up a store, people get killed; they blow up something else, people get killed.... They are crazy, they are whacky, they are bad people and we are going to make sure it happens as little as possible, but you can never claim victory because you always have a nut job someplace that's going to do something.[48]

Trump is not alone in this view: polls consistently reveal that the American public views mental illness as a common factor driving acts of violence, so it makes sense that this perception carries over to organizational fractures.[49] Researchers who study armed groups, however, tend to hold a different understanding. While certain personality traits might indeed drive people to engage in acts of violence, the groups themselves tend to behave quite rationally, leading to an assumption of *collective*, or organizational, rationality.[50] There are several reasons to believe this is correct.

For one, while not directly bearing on the topic of collective rationality, most studies find individual terrorists and militants to be rather ordinary. They are not crazed lunatics bent on destruction. Their backgrounds are often indistinguishable from ordinary citizens, undermining the idea that their groups will be similarly irrational.[51]

In addition, armed groups demonstrate remarkable strategic planning and foresight that attests to their collective rationality. Al-Qaeda's complex September 11th plot, involving 19 different hijackers, was first conceived in 1996—five years before the eventual operation.[52] Militant groups also respond in rational ways to changes in their strategic environment, demonstrating their situational awareness and analytic ability. Notably, there is strong evidence of a substitution effect in terrorist and insurgent tactics: when one approach is compromised, groups will shift to another mode of attack.[53] Scholars also find that groups vary tactics in accordance with public opinion and support for specific operations and targets. In the Israeli-Palestinian conflict, Hamas and Fatah moderated the number of suicide missions as support waxed and waned.[54] Together, these points underscore how armed groups make rather predictable and understandable decisions, conflicting with perceptions of their irrationality.[55]

Turning back to the topic of fragmentation, one can extend the logic of collective rationality to organizational splintering. Splits occur when members decide to exit an organization, and these subgroups should exhibit a similar rationality. Evidence supports this conclusion. As Jones and Libicki argue, "terrorist groups [that are about to splinter from their parent] conduct some form of implicit cost-benefit analysis" to determine if the effort is worthwhile and if they can successfully part ways.[56] If they believe the time is inopportune and they may fail, then they could delay their exit. Morrison similarly argues that a subgroup's decision to break away is a function of rational, internal calculations.[57] As he notes:

> The event of organisational exit and split takes place when one of the factions deems their presence within the terrorist organisation no longer tenable and that they are ready to move away from the parent organisation. This may be brought on by the completion of a satisfactory preparation for split whereby the dissidents believe that their future organisation has sufficient levels to support, membership and resources to both survive and be successful in their aims.

Membership, support, and sufficient resources are critical to the survival of splinter groups—indeed, to the survival of all armed groups—and the timing of a subgroup's departure is therefore carefully determined. Far from being irrational, this process is instead quite calculated.

In sum, there is a misconception that armed groups, particularly terrorists, are irrational and potentially unhinged. If this were true, studying organizational fractures and trying to predict their outcomes would be a futile endeavor. But research and historical patterns disprove this. There is indeed a method to their madness, and this should extend to how splinters emerge.

Myth 3: Splintering Is Abnormal

A third misconception held by a wide range of audiences—but particularly policymakers, practitioners, and researchers—is that splintering is an abnormal, dysfunctional organizational process. Perhaps it is something that well-managed groups can avoid. These events, however, are far from abnormal and are not necessarily undesirable either.

To begin with, groups of all kinds experience organizational splits. Hart and Van Vugt note that splits—or as they call them, fissions—occur in profit and nonprofit businesses, religious groups, political parties, nation-states, hunter-gatherer societies, and even groups of primates.[58] Regarding businesses, Dyck finds that nearly one in every five form by breaking away from existing firms.[59] Some of the emerging companies are quite successful, too: Puma split from Adidas, PayPal split from eBay, and AOL split from Time Warner. For firms, splits are a natural part of the business cycle, providing a critical opportunity for internal alignment and readjustment.[60] They also allow executives to separate out particularly effective or particularly problematic subunits, like GE did to its software business in 2018.[61]

Organizational splits are especially critical to the development of religious movements and political parties.[62] Seminal work by Stark and Bainbridge on religious schisms views them as natural processes that stem from internal disagreements.[63] "Schisms . . . are most likely to occur along lines of cleavage. That is, when internal conflicts break out in a religious organization, they usually do so between subnetworks that existed prior to the outbreak of dispute." And for political parties, an organizational split is often better than maintaining a united but heavily factionalized group.[k] With a few exceptions,[64] factionalism will "[undermine] party cohesion, effectiveness and management" while "policy debate in parties can become engulfed in factional squabbling and jostling."[65] A split can therefore improve party performance.

Data reveals that splintering among armed groups is just as common as it is among other organizations. For this project I collected new data on the origins of 300 randomly selected armed groups listed in the Global Terrorist Database, and I find that splinters emerged from approximately 28.3 percent of them. Other researchers uncover similar statistics. Among 181 armed groups in Africa, Braithwaite and Cunningham find that roughly one-third (66) formed by breaking away from a preexisting group—the most common pathway identified in the study.[66] The Minorities at Risk—Organization Behavior data set finds that just over 32 percent of groups splintered at some point, while it is closer to 37 percent for violent groups

---

[k] For armed groups, splintering is especially preferable to internal factionalism. While in other types of organizations factionalism might simply undermine coordination and cooperation, for armed groups it can be fatal. As I elaborate more fully in the following chapter, groups that are internally factionalized are prone to defection, leaking, and other deviant behavior. This is not simply problematic, but it can lead to death or imprisonment for all involved.

alone. Woldemariam finds that nearly a third of armed groups in sub-Saharan Africa have split,[67] and Lidow finds almost the same figure for groups operating in Liberia.[68]

Taken together, these studies demonstrate that splintering is far from abnormal. Instead, it is a common organizational process that affects groups of all kinds and that can sometimes serve an important function as well.

Myth 4: Splinter Groups Are Weak

A fourth misperception held by those who study and observe armed groups is that splintering indicates organizational weakness. According to this notion, groups that emerge from splintering should be short-lived and potentially weak, with little chance of success. In reality, there is little evidence to suggest that splinter groups should be universally weaker than others. By forming in this way, splinter groups may actually wield some significant advantages.

To begin with, groups that emerge from preexisting armed organizations should possess a high level of collective experience at the time of their formation. In other words, their members will tend to have more practical know-how than members of a group who are fighting for the first time. This might include information on how to organize, how to conduct certain types of attacks, how and from whom they can raise funds, and so on. This makes splinters distinct from nonsplinter groups and uniquely positioned to succeed, particularly in their formative years when disparities in levels of experience are most severe.[1]

Analogies from the business world underscore why collective experience matters. Studies find that the level of collective experience held by firm members upon inception is a crucial predictor of its long-term viability.[69] "Congenital experience," as it is often called, has a profound, positive impact on the likelihood of successful risk-taking and on the firm's expected profitability.[70] In a similar study of young firms' ability to internationalize their operations, Bruneel and coauthors write that:

> This type of congenital learning arising from the knowledge stock brought into a firm at founding through its founders' past experiences will have

---

[1] In other words, before the nonsplinter group has a chance to gain any experience.

an important imprinting effect on the firm's strategy. Previous actions and their outcomes are retained in the memory of the founders, resulting in interpretations and generalizations that can be drawn upon in decision making.[71]

Researchers have uncovered other benefits as well. For instance, experienced leaders not only bring their past experiences to bear on business operations, but they also bring with them well-developed social networks that firms can leverage to their benefit.[72] Firms can utilize existing networks to facilitate their expansion into new markets and to establish new business operations.[73]

It is easy to see how these same principles apply to new armed orga-nizations and specifically splinter groups. Since the odds of survival and success are already small, members' collective experience should make splinter groups both more capable and more likely to persist than similar nonsplinters. They can use their knowledge to more effectively organize and to anticipate potential challenges. These organizations can also leverage their members' existing social networks to secure funding, rally support, and gain recruits.[74] Of course, new groups that do not form by splintering also need to accomplish these same tasks to survive, but their lack of experience in these matters puts them at an immediate disadvantage.

From a tactical standpoint, previous experience provides armed groups with critical technical know-how. Two organizations illustrate this dynamic: the Red Brigades in Italy and the Real IRA in Northern Ireland. Before conducting violent operations, the Red Brigades had to first learn the basics of simply being a revolutionary organization. Their members, many of whom were students, had almost no idea how to manage a clandestine group and even less knowledge about how to coerce a state. Accordingly, some of their earliest efforts were reading books from revolutionaries in Latin America and sending operatives abroad for training. On the other hand, when the Real IRA formed by splintering from the Provisional IRA, it was able to immediately launch combat operations as its members had significant operational experience. Two of the Real IRA's founding members were Michael McKevitt, the former quartermaster general of the IRA whose specialty was arms procurement; and Frank McGuinness, a pseudonym for the IRA's deadliest and most notorious bomb-maker. The Real IRA's expe-rience contrasts sharply with the initial inexperience of the Red Brigades, demonstrating how the disparity in experience between splinter and nonsplinter groups might influence their operational patterns and their long-term viability.

Second, splinter groups are, by definition, formed by members of a pre-existing organization. This means that these individuals already know each other and have fought together, which builds trust and develops interpersonal bonds that foster cooperation.[75] For instance, consider the formation of the Irish National Liberation Army as it was still within the Official IRA: "Having been alienated from the Official IRA leadership for some time [INLA leader Seamus] Costello had anticipated his departure and in the final months deployed Official IRA personnel sympathetic to his ideas to carry out robberies to fund to new organisation."[76] The future members of the INLA started working together years before their eventual departure, both in terms of armed operations and as a voting bloc within the Official IRA, solidifying their commitment to one another. Extant research finds that this type of shared history and repeated interaction breeds trust, which is essential to armed groups that require secrecy and cooperation to survive.[m]

In sum, splinter groups are not inherently weak, unstable products of organizational chaos that are doomed from the start. They are actually uniquely capable organizations and their particular method of formation offers an advantage over their competition. Although they still encounter many of the same difficulties as nonsplinter groups—challenges of organizing, surviving, recruiting, and so on—their particular method of forming generally provides meaningful advantages in experience and and interpersonal trust. This underscores the importance of studying these groups more thoroughly.

---

[m] Why does trust matter and what role does it play in armed groups? Put simply, "When people trust each other transaction costs in economic activities are reduced, large organizations function better, governments are more efficient," and so on (Alberto Alesina and Eliana La Ferrara (2002). "Who Trusts Others?." *Journal of Public Economics* 85.2, pp. 207–234, p. 208). In other words, trust tends to improve organizational processes. In armed groups, when leaders do not trust their members to follow through with orders, they must expend additional resources to monitor their behavior, "[leading] to wasteful, inefficient resource allocation..."(Jacob N. Shapiro (2013). *The Terrorist's Dilemma: Managing Violent Covert Organizations.* Princeton University Press, p. 49). Resources can be scarce within armed groups, and failure to follow orders can lead to defection and insubordination that threaten the survival of the entire organization. As a result, "it is critical that (in addition to the requisite skills and knowledge) the soldiers of a unit build strong mutual bonds of trust and affection" (Laurence R. Iannaccone (2006). "The Market for Martyrs." *Interdisciplinary Journal of Research on Religion* 2.4, pp. 1–29, p. 11). Splinter groups should naturally form with even high levels of trust, offering a meaningful advantage.

Myth 5: Fragmenting Armed Groups Is a Good Counterinsurgency Strategy

If one believes that splintering is a sign of organizational weakness and internal discord, and that it produces weak and short-lived organizations, then it makes sense that causing organizations to splinter would be a beneficial counterterrorism and counterinsurgency strategy. This logic is commonly echoed by politicians. President George W. Bush remarked that the US was working to "break up [terrorist] cells."[77] President Barack Obama noted that US forces were "breaking up Al-Qaeda cells to disrupt their activities...."[78] And General David Petraus advocated for "[causing] divisions" among insurgents.[79] Understanding how armed groups splinter is therefore directly relevant to US foreign policy.

To begin with, much of the US Army's counterterrorist and counterinsurgency operations—their guide to defeating groups like Al-Qaeda and the Islamic State—is based on Field Manual 3-24 (FM 3-24).[80] At its core, the manual emphasizes a strategy to defeat irregular threats that is based on two overarching objectives: understanding the enemy and then undermining the enemy. Understanding the enemy is nothing new,[n] but their approach to undermining and defeating the enemy is novel. One of their main strategies is to essentially "divide and conquer": in effect, if one can fragment an armed group and split it apart, then it should become easier to defeat. The authors of FM-324 even note that "Rifts between insurgent leaders, if identified, can be exploited.... Offering amnesty or a seemingly generous compromise can also cause divisions within an insurgency and present opportunities to split or weaken it."[81] Here, we can see an explicit conflation between organizational splintering and weakness.

The divide and conquer approach has seemingly worked its way into counterterrorism and counterinsurgent strategies from its origins in interstate

---

[n] Carl von Clausewitz famously noted that the first step toward victory "is to identify the enemy's center of gravity" so that force can be concentrated on that point, crippling the adversary's abilities to resist (Carl Von Clausewitz (1982). *On War*. Penguin Publishing). Sun Tzu similarly notes that "If you know the enemy and know yourself, you need not fear the result of a hundred battles.... If you know neither the enemy nor yourself, you will succumb in every battle" (Sun Tzu (2010). *The Art of War*. Capstone Publishing.) More recently, counterinsurgent theorists like B. H. Liddell Hart, David Galula, and David Kilcullen all echo these same ideas, that intelligence, information, and a keen understanding of the enemy are essential to victory against irregular opponents (Basil Henry Liddell Hart (1967). *Strategy: The Indirect Approach*. Faber; David Galula (2006). *Counterinsurgency Warfare: Theory and Practice*. Greenwood Publishing Group; David Kilcullen (2010). *Counterinsurgency*. Oxford University Press).

warfare. When fighting other governments, states will target the command and control centers of their enemy's armed forces to prevent coordination and incite disorder. This was a core component of Germany's Blitzkrieg doctrine that made its military offensive so successful in World War II, and it was echoed even more recently when the United States invaded Iraq. Indeed, the logic of interstate conflict is such that a fragmented state or military is a weakened one; a fragmented state cannot bring its maximum power to bear on its adversary, and the balance of power thus swings in the opposite direction.[82]

Several problems arise when applying the logic of divide and conquer from interstate wars to subnational conflicts. First, a fragmented state's weaknesses are obvious: states are large, bureaucratic organizations whose warfighting capacity is a product of their coordination. Since modern combat units require extensive support and logistical networks to function effectively, disrupting rear operations significantly degrades their combat ability. Armed groups operate quite differently. Although certain groups like pre-9/11 Al-Qaeda, the Provisional Irish Republican Army, and the Tamil Tigers were highly institutionalized at different points in time, they do not depend on internal operations and logistics networks the way states do. Consequently, while splintering may prevent complex missions that rely on structural hierarchy,[83] it is unclear how a targeted fragmentation strategy should hasten a group's overall decline.

Second, a fractured organization may become a more dangerous and more capable organization in the future. Recently splintered groups are inherently, if only temporarily, weaker since manpower and resources have been divided, but with their battle-hardened members and previous training, they could evolve into increasingly capable and more resilient threats in the future. Setbacks like this are also critical moments for groups to reorganize and restructure, doing away with ineffective routines of the past.[84] Policies designed to fragment groups might therefore be trading short-term solutions for long-term uncertainty.

Third, although states may endeavor to fracture irregular opponents, not every case of organizational splintering should be seen as a military success; on the contrary, it may occur within a broader strategic logic to intentionally blunt the efficacy of government operations. Shapiro, for instance, argues that terrorist organizations will decentralize when facing government repression and infiltration, trading the ability to conduct complex missions for simple security.[85] Accordingly, FM 3-24 notes that ". . . a single insurgency may be in different phases in different parts of the country. Advanced insurgencies can rapidly shift, split, combine, or reorganize; they are dynamic and adaptive."[86] Thus, organizational splits might not only occur when groups are defeated or effectively repressed. They could instead signal that an irregular opponent is simply evolving.

Fourth, and finally, history has shown that the universe of splinter groups is highly diverse: some have gone on to be among the deadliest groups of the twentieth century while others have died out quickly. This uncertainty makes fragmentation an ill-advised strategy. The initial anecdotes cited in this book highlight a stark contrast between IRA splinters like the Provisional IRA, a group that was active for nearly 30 years, and those from the Popular Front for the Liberation of Palestine, each of which only managed a few attacks in their short lifespans. Before organizational splitting becomes a preferred counterinsurgency doctrine, we need to understand the factors that contribute to variations in fragmenting outcomes and identify the conditions under which fragmentation produces weaker, less durable, and less stable off-shoots.[87]

What We Actually Know about Organizational Splintering

Existing research on the splintering of armed groups has left many questions unaddressed. For instance, there is little work on how splits unfold, what motivates them, and even less on splintering as a counterinsurgency strategy. Instead, research mainly focuses on when and under what conditions splits occur. Scholars working in this area recognize that armed groups are products of the environments in which they operate and the pressures that they face. These pressures can exacerbate intragroup differences or allow subgroups to manipulate power dynamics, weakening the bonds that hold them together.

To this end, scholars find that both conciliation and repression affect organizational cohesion and influence whether splinters emerge. On the one hand, repression can sometimes make them more cohesive, but at other times, less cohesive. As Zimmerman wrote in 1980, "There are theoretical arguments for all conceivable basic relationships between government coercion and group protest and rebellion, except for no relationship."[88] In the cohesion-building camp are scholars like Simmel[89] and Coser[90] who argue that conflict can bind members of a group together out of fear, concern for their safety and security, and to collectively resist a threat. On the other hand, some researchers argue that violence makes rebel organizations more likely to fragment. State repression and the potential for personal harm can prompt fighters to reconsider their allegiance to the group.[91] Anything that raises this costs of collective mobilization, like state repression, can ultimately make groups less cohesive.[92] Aside from its effect on individuals, repression can also force armed groups to undertake structural changes—specifically, to decentralize—that leave them less able to prevent internal disputes from escalating into a rupture.[93]

Empirical research has produced mixed results. Asal, Brown, and Dalton[94] find that state repression is uncorrelated with the likelihood of internal schisms. McLaughlin and Pearlman, studying the Kurdish and Palestinian national movements, find "The effect of repression on movement unity is contingent on the preexisting consensus on a movement's institutional equilibrium."[95] If group members are satisfied with power and resource distribution, repression will make it stronger. When a group is not in equilibrium, however, the dissatisfied segments will capitalize on the uncertainty to establish institutional reforms that pose a challenge to vested organizational interests.[o]

While government interventions are important, what armed groups are doing also affects the odds that they split apart. Woldemariam finds that how well groups are faring in combat shapes their internal cohesion.[96] When groups are doing well, it is more likely that their fighters remain loyal.[97] When failure and territorial loss prevails, however, individuals will question their allegiance. In such cases, "losing territory prompts an organization's constituent units to question the cooperative bargain that is at the heart of the rebel organization. All things equal, fragmentation is more likely in such contexts."[98] Relatedly, Kenny[99] finds a relationship between burden-sharing and organizational cohesion. As he notes, "Disintegration was highest when there was a perception among the rank and file that commanders in the base areas were not sharing an equal portion of the burden of war."[100] These studies affirm that the probability of fragmentation is partly dynamic, being driven by a group's strategic choices and its interaction with the state.

Another factor contributing to organizational splintering comes from organizational characteristics that diminish control or that allow internal factionalism to take hold.[101] Research by Staniland, for instance, finds that

---

[o] McClauclin and Pearlman commendably move this discussion toward testable causal mechanisms and they avoid overanalyzing the conflict's "master cleavage."

when militants are assimilated into and build upon local institutions, they are more cohesive and less likely to "suffer from higher levels of internal feuding and disobedience."[102] Their cohesion makes it more likely for a cooperative, rather than pragmatic, leadership to form, and these organization should have a greater chance of developing robust internal institutions that prevent organizational fractures. Other structural characteristics like a factionalized leadership and decentralized authority should also increase the likelihood that an organization will splinter. With regards to leadership, Asal, Brown, and Dalton note that "Organizations with factional or competing leadership act to precipitate the organization into fission and schism because they allow for a plurality of potentially competing opinions, objectives, and priorities, and thus are more likely to break apart under external stressors."[103] Decentralized authority is expected to work through virtually the same mechanism: since the leadership has less control over the organization, it is easier for conflicts to develop and diverse opinions to take hold. When this happens, leaders under the decentralized authority have even less ability to hold the group together, punish defectors, and maintain control.[104]

Taken together, existing research tells us more about the conditions opportune for organizational splits, and less about their underlying causes or how they unfold. It is tempting to imagine how these conditions might also shape whether splinters will survive and how they will behave. For instance, perhaps splits occurring amid state repression will produce weak, short-lived breakaway groups and the parent's simultaneous demise. But this should be avoided: research from other domains suggests that the events contemporaneous to organizational splits offer few insights on their own.[105] As Finke and Scheitle write in the context of religious schisms, "We must be careful not to let the manifest drama of a schism distract us from its latent causes. Instead of looking to the surface phenomena that occur during schisms, we must examine the deeper sociological and organizational contexts that give rise to schisms."[106] Applied to armed groups, this analysis suggests that we can only learn so much from the "manifest drama" of repression, ceasefires, battlefield loss, and leadership decapitation. To understand these groups, we must instead dig deeper.

To illustrate the limited utility of focusing on external events when studying fragmentation, consider how government repression can precede organizational splits that set breakaway groups on fundamentally different trajectories. For one, state repression can inspire internal debate over a group's current strategies and tactics. Some may believe that the only way to respond and to demonstrate resolve is to escalate violence, conduct more attacks, and kill more civilians.[107] If the group's leadership disagrees, hardliners might break away to ramp up violence on their own. If the repressive intervention takes out a central figure in the group, leadership disputes might rupture the group along lines of personal allegiance. As new leaders vie for power, the group might splinter into competing factions that each claim legitimacy. And lastly, repression might cause a group's leaders to reconsider a recent conciliatory offer, making some fighters realize that they are unlikely to ever obtain a better deal. Accepting this deal, however, could prompt a spoiler to emerge in opposition.

While that example is theoretical, the experiences of armed groups in Pakistan facing the state's violent interventions provide a real-world illustration. For Lashkar-e-Jhangvi, the state's repression generated interpersonal feuds within the group that "centered around previously little-known figures in [their] hierarchy following the killing of its top leaders."[108] Fractures within Tehrik-e-Khudam-ul-Islam and Jamaat-al-Dawaw also occurred "along personality lines" following the death of influential commanders.[109] Within Lashkar-e-Taiba, meanwhile, the state's crackdown led to disagreements over matters of funding. The exact timing of their split—while one of the leaders was meeting financial supporters in Saudi Arabia—"indicates that Iqbal's policy of seeking more funds from the Arab countries," in response to the state repression, "was one of the stimuli for the split."[110] And finally, repression prompted some groups in Pakistan to increase the frequency and scale of their attacks, which eventually generated internal disagreements over tactics and the acceptability of indiscriminate targeting. Describing his rationale for departing the Afghan Taliban, one Jamaat-ul-Ahrar commander noted that "[They] have no vision and mission except to kill innocent people."[111] As these examples demonstrate, focusing exclusively on external events neglects the very different internal dynamics motivating these splits—dynamics that fundamentally affect how the process will unfold and how the emerging splinter group will behave.

The Argument, in Brief

When armed groups fragment, why do some splinter groups radicalize, and why do some survive while others fail? My theory is simple: to understand the trajectory of splinter groups, we must first understand how and why they break away.

When it comes to radicalization, armed groups break apart for many reasons but most come down to a combination of three different causes: strategy, ideology, and/or leadership. These disputes influence which members join the splinter and which ones stay behind with the parent. Arguments over strategy should be the only disputes that draw hardliners—the most radical members—into breakaway splinters, and organizations with more violent hardliners will veer towards more violent operational profiles.[P] Disputes over leadership or ideology, meanwhile, will not necessarily appeal to hardliners who advocate for the use of force.

When it comes to the survival of splinter groups, splits can occur in one of two ways: unidimensionally, when the group is divided over a single issue (like strategy); or multidimensionally, when the group is divided over two or more issues (like strategy and ideology, or leadership and ideology). Unidimensional splits tend to produce more resilient breakaway groups. This is because single, shared disagreements lead breakaway groups to attract more homogeneous subsets of members, facilitating cooperation, cohesion, and organizational security. Multidimensional splits can attract members with competing views of their organizational future, raising the risks of internal disagreement, defection, and insubordination. These unfavorable dynamics can quickly lead to a splinter's demise.

Taken together, splinters breaking from their parent organizations over strategic differences are the most likely to radicalize, while those emerging unidimensionally are the most likely to survive. This discovery is significant because it shifts attention away from scholars' overwhelming emphasis on external events to explain the breakdown of armed groups. Instead, by looking inside these organizations, we can better explain variation in their breakdown as well as variation in the new groups that emerge. External events obviously matter a great deal: state repression can prompt or hasten internal disagreements, but it cannot explain why groups eventually split or how breakaway groups will eventually behave.

[P] Disputes over strategic moderation often occur, but they rarely lead to the formation of new armed organizations. Rather, they often precipitate defection: either to the government, to other groups, or away from violence. I discuss this more in the following chapter.

What about Different Types of Armed Groups?

One might note that my discussion so far has not focused on one particular type of organization. Instead, I use terms like "militant organization," "armed group," and "violent nonstate actor" interchangeably. Why not discuss these dynamics as they pertain to terrorists, rebels, insurgents, or guerrillas? The reason is that organizational dynamics cut across these classifications and operate regardless of how armed groups employ violence, and my argument is that these organizational dynamics are critical to explaining how splits unfold and how splinter groups behave.

To begin with, it is certainly not uncommon for researchers and especially policymakers to focus on particular types of armed groups. The Department of Homeland Security infamously created the color-coded National Terrorism Advisory System. There is also the National Counterterrorism Center, the Bureau of Counterterrorism and Counter Extremism at the State Department, and the Joint Terrorism Task Force at the FBI. The US Army, meanwhile, famously adopted Field Manual 3-24 that is simply titled "Counterinsurgency."

There are compelling reasons for governments to classify and distinguish between types of armed groups. Apart from the political motivations for labeling a foreign actor as a "terrorist"—which can be significant—different types of armed groups will pose different types of threats. Terrorist groups like the Ku Klux Klan will generally attack a different set of targets and with different methods than an insurgent group like the Taliban. Accordingly, different armed groups will require a distinct set of defensive measures and offensive tools.[q]

Researchers sometimes make similar distinctions when studying phenomena that are germane to specific types of organizations or armed conflicts. For instance, many aspects of civil wars and terrorist campaigns are fundamentally different, and combining both into a single study of subnational conflict might not make sense. Civil wars may exhibit competing territorial control and relatively conventional battles,[112] while terrorist groups rarely hold territory and usually attack soft targets to coerce foreign

---

[q] Classifying armed groups is also important to the application of domestic and international law, especially relating to terrorism.

governments. When studying the strategies of violence, then, it would make sense to differentiate between types of armed groups.

In other cases, as with studying organizational dynamics, the distinction between different types of conflicts and armed groups serves little purpose. This is because at their cores, terrorists, insurgents, and civil war rebels are groups of men and women seeking to affect social, political, or economic change through the use or threat of force. As groups of people, organizational dynamics are omnipresent and virtually all groups will experience some of the same challenges and constraints regardless of the circumstance in which they find themselves. Most people will be familiar with some of these. For instance, just recall a time when a group assignment went poorly: maybe the issue was ineffectual or nonexistent leadership, competing leadership, unmotivated peers, a lack of familiarity and trust between group members, or an uneven distribution of labor. These dynamics are essential to understanding how a group performs regardless of whether their task is a classroom assignment or a terrorist operation.[r] By utilizing this perspective, and by drawing upon our vast knowledge of organizational dynamics in more familiar settings, researchers have made significant progress towards understanding how armed groups function.[113]

Two areas of research demonstrate the utility in studying the organizational dynamics of armed groups. First, researchers find that organizational structures impact how organizations of all kinds behave. This is just as true for business firms as it is for armed groups. For the latter, structures often conform to one of two ideal-type categories: either hierarchical or decentralized (also called networked).[114] Each has pros and cons. On the one hand, hierarchical structures are better for managing operations, delegating responsibilities, and controlling violence—all of which facilitate the planning and execution of increasingly complex missions—but they are easier to infiltrate.[115] On the other hand, decentralized structures are harder to identify, infiltrate, and destabilize, but they make complex missions more difficult.[116] Kilberg also finds that organizational structures affect target

---

[r] Some might find it hard to believe that we can find useful analogies between insurgent groups seeking to overthrow a government and business firms seeking to make a profit, for instance. In many ways the use of violence and the clandestine nature of militant groups will complicate any comparison. But as Shapiro notes, "The willingness to kill civilians as a legitimate means to a political end is certainly radical, but terrorists are every bit as, if not more, venal, self-important, and short-sighted as the rest of us" (Shapiro, 2013, p. 2). In effect, groups of all kinds exhibit remarkable similarities because they are all comprised of humans who cannot set aside their most basic characteristics.

selection, with centralized groups attacking a greater proportion of hard targets (e.g. government and military installations) compared to soft targets (e.g. civilian markets).[117] By understanding how armed groups are structured, then, we can better understand their operational profiles and even their resilience.

Second, scholars have increasingly focused on how an armed group's organizational capacity (or capital) influences its ability to innovate. As Horowitz writes, organizational capital:

> refers to the previously intangible aspects of organizational strength that firms draw upon when facing periods of industry transition. . . . Organizations with a high degree of organizational capital are much better able to take advantage of new innovations and transform themselves successfully for the future than organizations with a low degree of organizational capital.[118]

Traditionally, economists used organizational capacity to explain variation in how effectively business firms adapt to innovations like new technology (e.g. computers) or new production methods.[119] With regard to armed groups, organizational capacity operates similarly. It shapes their ability to employ innovative tactics and strategies, particularly those that disrupt existing practices. Specifically, Horowitz finds that younger terrorist groups—with the greatest capital—are most likely to adopt suicide bombing since they can easily adapt to new operational needs and routines. Older groups with entrenched practices and less organizational capital were more resistant to tactical change.

Taken together, organizational structure and organizational capacity are two dynamics that affect groups of all kinds. As innate characteristics of human groups, their influence can be seen among terrorists and rebels, business firms and political parties. My theory of how splinters emerge and subsequently behave is similar. By drawing upon common organizational and even human dynamics to understand their trajectory, it will help make sense of a wide array of armed organizations.

Looking Ahead

The rest of this book proceeds as follows. In Chapter 2, I present my theory of how armed groups divide and how splinter groups behave. Drawing upon research from fields as varied as organizational ecology and religious studies, I show how and why the motivations underlying organizational breakdown have enduring effects on the new groups that emerge. It comes down to membership: understanding the disputes that precede organizational ruptures are critical to explaining why the membership of a breakaway group takes the form that it does. Once we understand a group's membership dynamics, we can better understand its trajectory and behavior. At the end of Chapter 2, I present a brief case study of a major split among the Basque terrorist group ETA that produced two distinct organizations: ETA-*militar* and ETA-*político-militar*. This demonstrates the practical and analytical utility of my theory, and it provides readers an opportunity to see how group breakdown in the real world follows closely to my theoretical model.

In Chapter 3, I test my theory qualitatively. I present a case study of fragmentation among armed groups in Northern Ireland that evaluates the mechanisms and demonstrates the applicability of my theory. This chapter begins with an overview of the conflict between Republican and Loyalist dissidents as well as a brief timeline of events. I then move on to the primary focus: an investigation into the trajectories of two splinter groups, the Irish National Liberation Army (INLA) and the Real Irish Republican Army (RIRA). I trace the formation of both groups from their development as subgroups within the Official and Provisional Irish Republican Armies, respectively. I show how their formation corresponds to my model of group breakdown and how their disagreements shaped their long-term organizational trajectories. This research draws upon internal group documents, formerly-confidential government assessments, and contemporaneous news articles and interviews that I uncovered during months of research in Belfast, Dublin, and London.

In Chapter 4, I test my theory quantitatively. I begin by describing the new data set created specifically for this project, and I then present descriptive trends about the frequency and other characteristics of organizational fractures. For instance, I evaluate the average age at which groups split, whether all types of organizations are equally likely to split, and other differences between splinter and nonsplinter organizations. Then, I use the newly collected data to study the radicalization of breakaway groups. I look at several different metrics: the frequency of an armed group's attacks, the total number of fatalities from its attacks, the average lethality of individual attacks, and the adoption of suicide bombings. These results show that not all splinter groups are created equal, and that—as my theory predicts—those emerging from strategic disputes are especially deadly. Next, I evaluate the survival of splinter groups. I find significant differences in how long they persist, with unidimensional-strategic splits producing groups that are nearly 20 percent more likely to survive past their first year than others. On the other hand, splinters emerging multidimensionally and over leadership disputes in particular tend to die out at significantly quicker rates. This difference only widens with time, and at ten years after a split, unidimensional splinters are nearly 35 percent more likely to endure. Taken together, these statistical tests demonstrate a strong link between how and why splinters form and their ultimate trajectories.

In Chapter 5, I study a relevant but atypical case of organizational splintering: that which produced the Islamic State from the ranks of Al-Qaeda. Unlike the previous case study where I carefully test the mechanisms at work, here I evaluate how my theory applies and what it reveals about a modern organizational rupture with contested information and an unusual trajectory. Leaders of Al-Qaeda effectively imported a faction in 2004 when they allied with a new group, led by Abu Musab Al-Zarqwawi, which would soon become Al-Qaeda in Iraq. This faction held a distinct ideology, strategy, and overall organizational ethos. So when the split occurred, it was this preexisting faction that effectively broke away. Nonetheless, my theoretical framework reveals how the international community should have been aware of this potential split and aware of how dangerous this new entity could become. The Islamic State's disagreements with Al-Qaeda's leaders were no secret; accordingly, it was obvious how they would behave once independent. As I anticipate, internal politics, and not solely counterterrorist interventions by the US and others, were critical to how these events unfolded and how these groups subsequently evolved.

Finally, in Chapter 6—the Conclusion—I discuss the implications of my research. I show how my theory and findings help us understand the formation of armed groups, the trajectory of fragmented conflicts, and the efficacy of counterterrorism and counterinsurgency strategies. As for the latter, I discuss how states can foment organizational ruptures that create weak, short-lived organizations while avoiding ruptures that might create more formidable opponents. I then consider the fate of splinter groups' parent organizations. While not my primary focus, I discuss why some splits may precipitate a parent organization's demise while others might leave them relatively unscathed. Finally, I extend my findings beyond armed groups. I show how this research sheds light on the breakdown of nonviolent movements and how it is especially useful for understanding the development of radical flanks.

2

How Armed Groups Divide

Nearly 30 percent of all armed groups form by breaking away from preexisting organizations. But why do some of these new groups radicalize while others do not? And why do some survive while others fail? I argue that the answer to these questions lies within the nuances of how organizational ruptures occur. It is important to understand this process and not, contrary to most existing work, black-box the dynamics that culminate in an armed group's rupture. By focusing solely on the high-profile and often dramatic moment when organizational splits occur, one ignores everything that precedes and even causes these critical events. It is only by understanding and disaggregating this process that we can identify when and where variation is introduced and why splinters set off on different trajectories. As I show, variation in how splits unfold shapes who joins the emerging splinter and who stays behind with the parent. We can use this insight to predict their odds of both radicalization and survival.

In this chapter, I present my model of how armed groups divide. It occurs in three stages: in Stage One, internal unity prevails. Organizations strive for this kind of unity to maintain cohesion, reduce security risks, and to operate most efficiently. In Stage Two, factions (or subgroups, which I use interchangeably) form as members coalesce around their shared dissatisfaction with the status quo. In Stage Three, factions break away to create new armed groups that rectify their grievances and fulfill their organizational visions. Ultimately, variation in Stages One and Two can help explain rates of radicalization and survival among the groups that emerge in Stage Three.

When it comes to radicalization, some might assume that all armed groups are relatively *radicalized* already. However, there is actually significant variation in their strategic and operational profiles. For instance, armed groups can vary according to: whether or not they kill civilians; whether they conduct indiscriminate or targeted attacks; whether they attack predominantly soft (e.g. civilian) or hard (e.g. military and police) targets; their frequency

Divided Not Conquered: How Rebels Fracture and Splinters Behave. Evan Perkoski, Oxford University Press.
© Oxford University Press 2022. DOI: 10.1093/oso/9780197627075.003.0002

of attacks and even their choice of tactics (e.g. suicide bombings versus conventional explosives).[a]

I argue that variation in radicalization is strongly linked to the disagreements motivating subgroups to form (in Stage Two) and splinters to emerge (in Stage Three). When splinters break with their parent over strategic disputes—disputes that most commonly occur over the desire to expand or renew the use of force—they generally appeal to hardline operatives who prefer more radical operational profiles. Splits over ideology and leadership, on the other hand, appeal to different subsets of fighters who may be uninterested in escalation. In effect, disputes over strategy should produce highly radicalized splinters.

Just as there is significant variation in rates of radicalization, armed groups also survive to very different extents. Many of the most notorious organizations that have persisted for decades—like Al-Qaeda, the Taliban, the Ku Klux Klan, and others—are actually outliers since most armed groups typically fail, or die out, within the first few years. In fact, nearly half of all terrorist organizations only last a single year.[1]

I argue that the survivability of splinter groups is linked to the breadth of disagreements that initially motivate factions (Stage Two) to coalesce within armed organizations. Subgroups that develop around multiple grievances, in what I term a multidimensional split, broaden their appeal to a more diverse subset of members that hold different perspectives on how the new group should organize and behave. When a single, shared issue divides the parent group—in a unidimensional split—then the breakaway group attracts a more homogeneous subset of fighters with similar preferences. A single, polarizing grievance is a clear indication of the new group's ambitions, and it cultivates a more cooperative, like-minded base of recruits that make the splinter's long-term survival more likely.

Consequently, my model of organizational splintering implies that the dynamics of the split itself—dynamics that take place below the surface, before any formal split even occurs—shape breakaway groups in important ways. When the preferences of departing members are aligned, survival becomes more likely; and when their preferences are for greater radicalization, extremist behavior is expected. The causes of organizational splits

---

[a] To illustrate, consider two at *very* opposite ends of the spectrum: the Earth Liberation Front (ELF) and Hamas. According to the Global Terrorism Database, ELF has conducted 89 attacks, with 77 against facility or infrastructure targets, and has killed zero people. Hamas, meanwhile, has conducted 440 attacks. 175 have been against private citizens, and several hundred civilians have been killed.

and the movement of members into breakaway organizations is therefore critical to understanding how these processes will unfold and how splinters will evolve.

The Process of Organizational Splitting

The breakdown of an armed group occurs in three stages: Stage One, when an organization enjoys high levels of internal unity; Stage Two, when internal disagreements take hold and subgroups (factions) develop around things like competing leaders or different strategies; and Stage Three, when a faction breaks away to become independent.

In fact, schisms within armed organizations follow relatively logical processes that will likely seem familiar. Many readers will implicitly recognize this pattern, moving from relative unity to internal disagreement to splinter formation, as it is common to organizations of all kinds: from business firms to religious sects.[2] Regarding how terrorist groups emerge from schisms within political parties, for instance, Weinberg and Eubank note that:

> The other major pathway leading from party politics to terrorism is one where a factional division occurs within a party the result of which is the faction's defection for purposes of conducting terrorist operations.... Extremist parties are frequently susceptible to factionalism based on conflicting interpretations of revered concepts as well as differing assessments of how political doctrines should be read in light of current developments. In many cases the result of these disputes is a new party emerging from a faction of the old one.[3]

Compare that to a description by Zuckerman about a schism within a Jewish community group:

> In the early 1990s, the single center of Jewish communal life in Northweston, the Northwest Temple, experienced in-fighting, factional division, and an eventual schism. What had once been the sole congregation in Northweston, uniting this small Jewish community "under one roof" since the 1950s, suddenly experienced a hostile break-up which resulted in the establishment of a separate alternative congregation, the Northwest Minyan.[4]

Although they take place in fundamentally different settings, these brief examples reveal some important facts about organizational ruptures. They are not haphazard, spontaneous, or unintentional. Instead, they are usually preceded by a period of internal tension and factionalism where people with similar ideas group together. This provides observers and even group members with evidence that a rupture may be coming. Moreover, the factions that ultimately form are organized around their shared dissatisfaction with the status quo and their competing visions for the organization and its future. Logically, we can use this information to understand how factions will behave once a rupture occurs and they become independent.

This same process applies to splits within armed groups. Although they embrace violence and kill civilians, rebels, terrorists, and insurgents are still, at their cores, groups of people whose behavior is shaped by social and organizational forces. They cannot escape this fact, and it means that their organizational ruptures follow a well-understood, predictable pattern of events that is not dissimilar to splits in other types of groups. In the following pages, I detail this process more carefully and I show how we can leverage it to better understand what happens after armed groups divide.

Stage One: Organizational Unity

The first step towards an organizational split is for factions to develop, but in Stage One, they have not yet emerged. This is depicted in Figure 2.1. At this point armed groups experience relative internal unity, they functions as cooperative entities, and they are free of cohesive factions or subgroups. There may still be specialized units or administrative bodies (e.g. leadership councils and operational or regional cells) but these are fundamentally different from the factions that break away to become splinter groups.

So what exactly are factions in this context? Borrowing from seminal research by Zariski[5] on political parties,[6] they are "any intra-party combination, clique, or grouping whose members share a sense of common identity and common purpose and are organized to act collectively—as a distinct bloc within the party—to achieve their goals." Their common purpose is especially important: it is what binds them together and what motivates them to "compete for the acquisition of influence" over the party, its institutions, and its decisions.[7] In addition to their shared desire for change that lies at the core of why factions exist, another important characteristic is that

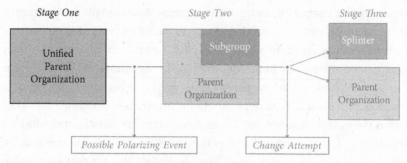

Fig. 2.1 Stage One of an organizational split

they are informal. Their boundaries and membership are not clearly defined; individual or collective leadership may develop but it lacks proper institutionalization;[8] and most importantly, those involved in the faction still view themselves primarily as members of their present organization. Put another way, their allegiance is still with the broader group they are working to change, and not solely with their factional compatriots. As factions continue to develop, however, their allegiance slowly shifts to the faction itself.

Having defined subgroups, recall that in Stage One of organizational splits, these units do not yet exist. Not all armed organizations manage to achieve this level of internal unity, and some will begin their life-cycles at Stage Two with factions already present.[b] For instance, armed groups that previously operated in other capacities, like as a political party,[9] may be encumbered by intraorganizational divisions and loyalties upon forming. This is precisely what occurred in Syria around 2016 when several factions (Harakat Nour Al-Din Al-Zinki, Liwa Al-Haq, Jabhat Ansar Al-Din, and Jaysh Al-Sunna, and others) merged together to form Hay'at Tahir Al-Sham.[10] Members of the new group brought along their preexisting factions and loyalties, so it is no surprise that infighting soon began.[11]

Armed groups are especially likely to experience the unity of Stage One in two cases. First, it is more likely in young organizations where the social bonds that underpin subgroups have yet to form.[12] It takes time for individuals to identify each other, their common grievances, and then aggregate into cohesive subgroups. In new organizations, there is simply less time for this process to occur. In addition, group members might also be more supportive of their leaders during an organization's early stages, opting

---

[b] This is important because it affects the odds—but not the motivations—for splintering to occur.

to give them a chance before openly dissenting. Accordingly, data presented in the coming chapters reveal that groups in their first years of existence typically experience fewer organizational splits. Second, groups are more likely to experience Stage One in the aftermath of an organizational split, whether they are the parent or the breakaway splinter. This is because organizational fractures provide a critical opportunity for preference realignment.[13] Splits may serve as release valves for intraorganizational tension by providing an opportunity for dissatisfied members to depart and to join an organization more in line with their own views. The observable effect is that groups that have recently split are less likely to be comprised of subgroups that differ substantially in their views.

Stage Two: Factions Form

The first sign of a possible organizational split is when factions develop in Stage Two, depicted in Figure 2.2. This is difficult for armed groups to avoid since factionalism occurs within all kinds of organizations. There are several reasons why.

First, innate human characteristics are partly to blame.[14] Numerous studies reveal that individuals commonly take actions to reduce their feelings of uncertainty. As Hogg skillfully notes:

People need to feel certain about their world and their place within it; certainty renders existence meaningful and gives one confidence in how to behave, and what to expect from one's physical and social environment. Uncertainty about one's attitudes, beliefs, feelings and perceptions, as well as about oneself and other people, is aversive because it is ultimately

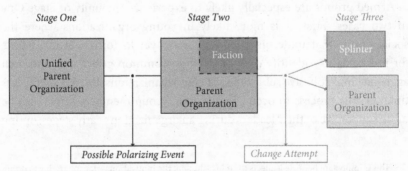

Fig. 2.2 Stages One and Two of an organizational split

associated with reduced control over one's life. For example, Kramer and Wei (1999) discuss how uncertainty about one's relationship with others in a group may provoke mistrust and paranoia. Thus, uncertainty motivates behavior that reduces subjective uncertainty.

In large groups, uncertainty can prompt members to seek out others who feel the same way to confirm that they are not alone in holding a given set of ideas. Even outside of group settings, this sort of behavior can be observed among individuals who gravitate toward like-minded communities online, in person, or via their news choices. The natural tendency toward confirmation bias is a good example: people tend to believe information more readily if it comports with their preexisting opinions, providing them with evidence that their ideas are correct.[15] All of these actions aim to reduce uncertainty by finding a like-minded community, and these same dynamics unfold within armed groups when members begin to question their organizational trajectories.[16]

Second, factions tend to form in most organizations due to the underlying diversity of group members, including diversity based on gender, age, occupation (or role), and ethnicity. According to theories of self-categorization and social identity—which suggest that humans are social, group-minded individuals who sort themselves into categories based on underlying attributes[17]—these basic distinctions can serve as the building blocks for factions to emerge.[18] In other words, when people are part of a large group, they might seek out others who share their language or are from the same hometown, and these connections can foster discussion that eventually sets the stage for a more cohesive faction to emerge around shared dissatisfaction.

Third, external events—what I call *polarizing events*—can also prompt factions to form. While human nature and uncertainty are sufficient, external events can accelerate factions by creating friction between members, generating even more uncertainty, and instilling a sense of urgency.[19] In their study of Czech political parties, for instance, Shriver and Messer find that external forces were the most significant factor underlying the development of factions. "Cleavages emerged . . . as activists developed competing strategies for responding to the contraction of political opportunities" that was instigated by forces beyond their control.[20] Likewise, factions emerged in the Nationalist Party of China with the death of Chang Ching-kuo,[21] and the Austalian Labor Party factionalized when a new competitor emerged, the National Centre Left.[22] Among each of these political parties, the polarizing event in question generated uncertainty and urgency, and factions soon developed around competing responses. Similar processes affect armed groups. A negotiated settlement with the state could create confusion as members consider their future and whether any settlement could truly achieve their objectives. Or, the death of a key leader could galvanize supporters around competing heirs. This recently occurred with Al-Qaeda following the death of Hamza Bin Laden, with many speculating that a split could soon occur.[23]

As individuals naturally seek out peers who look and think similarly, factions will slowly develop within organizations. These factions eventually, after repeated interaction, "...coalesce into a subgroup that [possesses] a true group identity."[24] What lies at the core of these "social subgroups" is their "similar attitudes and enduring beliefs,"[25] or, as Zariski[26] notes, their "sense of common identity and common purpose" that is usually centered on a desire, or a shared preference, to affect organizational change.[c]

One may be wondering if it is even possible to observe factions at this point and before they break away. If not, evaluating where groups stand along this spectrum of splinter group formation would be difficult, and anticipating the emergence—let alone the trajectory—of breakaway groups would be even harder. Fortunately, it is possible when one looks carefully for signs of intra-group tension and for the formation of distinct blocs. Consider the Islamic State (IS). Although IS split from Al-Qaeda in 2014, keen observers could anticipate this rupture well in advance. The disagreements between Ayman Al-Zawahiri (the leader of Al-Qaeda) and Abu Musab Al-Zarqawi (the leader of Al-Qaeda in Iraq) during the mid-2000s were widely understood at the time. In one especially significant moment, US forces intercepted a letter from Zawahiri to Al-Zarqawi discussing their tactical and strategic choices. In it, "Zawahiri expresses total agreement with the goals of the jihadist military efforts in Iraq but expresses grave reservations with [Zarqawi's] tactics."[27] Zawahiri felt that IS was undermining their success by indiscriminately killing civilians, be they Muslim or not. Al-Qaeda, in contrast, believed in a more ground-up and (relatively) hearts-and-minds campaign to cultivate local support. Even though the formal

---

[c] Here, it is important to note what "preferences" mean in this context. Researchers like Shapiro (2013, p. 29) rightly differentiate between induced and underlying preferences. In effect, underlying preferences represent what one prefers in the abstract, whereas induced preferences represent those preferences in light of prevailing information and interpretations of that information. For my purposes, however, I am primarily interested in the preferences members hold regarding their organizational future, which is a specific form of induced preferences.

split between IS and Al-Qaeda did not happen until 2014, this intercepted document gave US forces insight into the tight-knit faction operating around Zarqawi[28] and the growing rift between Zarqawi and Zawahiri.[29]

Factions in Armed Groups

While factions can form in almost any organization, they are especially likely to form within armed groups for two reasons.

First, because of the potential gravity of one's uncertainty. While uncertainty in a political party might correspond to policy failure or losing reelection, uncertainty for someone in an insurgent group can be a matter of life and death. This constant pressure tears at organizational unity and makes the task of finding affirmation in one's peers even more compelling. As such, it is not surprising that "Scholars who have done extensive interview work with terrorists report their organizations are torn by strife and disagreement."[30] When it comes to matters of life and death, passions are easily inflamed.

Second, research finds that groups with increasingly exclusive membership criteria tend to experience greater levels of internal factionalism and produce more splinter groups. Exclusive membership means that an organization views participants as "zero-sum," such that multiple memberships are not permitted.[31] This is precisely how membership works in an armed group; membership in *other* armed groups is strictly prohibited, and for some, being a member is a full-time occupation that supersedes commitments to friends and family. One hypothesis is that membership in exclusive organizations is especially stressful and the exclusivity permits few options for dealing with internal disagreement; it is either conform or exit. As Zald and Ash add, "The exclusive organization not only requires that a greater amount of energy and time be spent in movement affairs, but it more extensively permeates all sections of the member's life, including activities with nonmembers."[32] This can produce feelings of alienation and stress that further undermine organizational unity.[33]

Stage Three: Faction Exit

The final act of an organizational split is the most dramatic, the most chaotic, and also the most visible: when the faction breaks away, severing ties with its parent to form a new, independent armed group of its own. This is depicted in Figure 2.3. It is not surprising that this is where most research on group

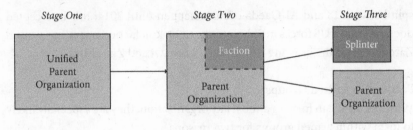

Fig. 2.3 Stages One, Two, and Three of an organizational split

breakdown is focused since it is an easily observable event that often attracts media coverage and incites international debate.

To reiterate, in Stage One of the splintering process, the armed group is free of cohesive factions; in Stage Two, factions begin to form around shared preferences and dissatisfaction with their present organizational trajectory; and finally, in Stage Three, the faction breaks away. It departs the parent organization for a simple reason: "to establish a new group more in line with the viewpoints and expectations of that section of the membership."[34] In other words, they are setting off on their own to fulfill their ambitions and to become the group they have envisioned. So in addition to determining who joins the faction in Stage Two, the faction's motivations for breaking away also influence how the subgroup will look and behave once independent.[d]

Why do factions break away when they do? To begin with, the existence of factions does not *necessarily* lead to an organizational rupture.[35] Of course, factions undermine cohesion and efficiency, but a group can survive even with relatively well-defined factions operating within its midst. Perhaps they can cooperate, or perhaps leaders can devolve authority to let them act semi-independelty. Decentralization was the preferred option for the Movement for the Emancipation of the Niger Delta, in Nigeria, a group consisting of numerous factions without a clear, single leader.[36] Subgroups might also

---

[d] There is, of course, some stochasticity to this rational model of organizational breakdown. Not every individual departing with the breakaway group has rationally evaluated the positions of the splinter vis-a-vis the parent. Some may join the breakaway group due to personal connections—for instance, being friends with others in the faction or even the faction's leaders. Nonetheless, there is strong evidence to suggest that personal connections are secondary to the underlying grievances. Describing the split between the Provisional IRA and the Official IRA, for instance, a former member noted that "The choice between the Provisional and the Officials was a military decision on how things should go" (Lorenzo Bosi and Donatella Della Porta (2012). "Micro-Mobilization into Armed Groups: Ideological, Instrumental and Solidaristic Paths." *Qualitative Sociology* 35.4, pp. 361–383, p. 370).

decide that, at least for the time being, staying in place is safer or more beneficial than breaking away.

There are other ways that splits can be avoided. Organizational ruptures usually follow some *change attempt* where members of the faction seek to address their grievances and enact change—or "apply their voice."[37] This form of subgroup lobbying is not uncommon, and it can be successful since both sides recognize the benefits of "split avoidance."[38] For the splinter, the desire to avoid the uncertainty of independence cannot be overstated. Faction leaders contemplating a split have an incredibly difficult task ahead of them: challenges of fund-raising, recruiting, organizing, establishing foreign connections, and preparing for potential intergroup conflict. The high failure rate for nascent armed groups attests to the seriousness of these challenges.[39] If splits can be avoided and grievances addressed, many would-be splinters will welcome the opportunity. This is precisely what happened in Northern Ireland when the Irish National Liberation Army (INLA) was forming within the Official IRA. One member of the INLA faction noted that their leader, Seamus Costello, "wanted to win out through that structure. The last thing I think he wanted to do was to leave the officials because to splinter weakened your position."[40]

For their part, leaders of factionalized armed groups also have options available to avoid or delay splits. They, too, usually hope to avoid splits that rob them of valuable fighters and that generate new competitors. If leaders recognize that a faction exists and disagrees with the group's current trajectory, they can take steps to accommodate the faction and change their behavior. Accommodation commonly occurs with hardline factions that want to increase the scale or frequency of attacks. Fearing the repercussions of ignoring their demands, leaders will escalate the organization's operations as a sort of concession. Hamas, a terrorist group in Palestine, has been known to grapple with this very problem, purportedly conducting certain attacks specifically to appease internal factions and to maintain unity.[41] A similar dynamic has repeatedly plagued negotiations between the US, Afghanistan, and the Taliban, with the latter rejecting deals that would principally anger its most hardline elements.[42] Leaders of armed groups can also implement more forceful strategies to avoid ruptures and quell dissent. If accommodation seems impossible—perhaps when members disagree over an already accepted peace offer, or more fundamentally with who is in control—leaders might sanction or even expel members[e] of an internal faction. In 2003, the Pakistani-based group known as Jaysh-i-Muhammad expelled more than a dozen members for conducting unsanctioned attacks.[43] Again, this demonstrates why understanding a faction's motivations has important consequences—not only for understanding how it eventually behaves but also because it sheds light on whether and how a group can resolve its internal tensions.

Finally, external events can also intervene to halt organizational splits, even though they are usually studied for their split-inducing properties. Indeed, this is not a process that solely runs in one direction, but it can be reversed. Sometimes, a "harmonizing event" can compel "all members [to come] together and [declare] their commitment to maintaining a unified organization."[44] This could be a dramatic increase in the rate of governmental repression or extrajudicial killings, or perhaps an attack gone spectacularly wrong. In Northern Ireland, for instance, the Real Irish Republican Army's disastrous bombing at Omagh inspired two other groups with historically discordant intragroup relations—the Irish National Liberation Army and the Continuity Irish Republican Army—to unilaterally accept ceasefires.[45] Another harmonizing event could be the death of a beloved leader. Analysts feared that Osama Bin Laden's death in 2011 could potentially unifying Al-Qaeda. Though no longer central to their day-to-day operations, his death could spark a renewed Al-Qaeda offensive as members lashed out to avenge his death. A communique from Al-Qaeda only a few days after Bin Laden's death implored members to do just that, while also using the occasion to call for unity among its ranks.[46] Of course, these very same events could have the opposite effect and accelerate fractures, underscoring how difficult it is to make assumptions about the likelihood of organizational splits from external events.

Ultimately, when mechanisms of split avoidance fail,[f] a faction may decide that its position within the organization is no longer tenable.[47] Its members will calculate that breaking away is worthwhile and they exit the group to start anew.[g] Research suggests that this decision is not taken lightly and

---

[e] While expulsion might simply hasten a faction's pre-planned exit, the faction could also not be ready or willing to depart. This could prevent it from realizing its potential as a new armed group.

[f] For instance, perhaps no mutually agreeable solution exists between group leaders and the subgroup.

[g] For this reason, it is difficult to predict the exact timing of an organizational rupture as it is effectively idiosyncratic.

is driven by both conflict and intraorganizational dynamics.[48] Subgroup leaders may bide their time until the parent group holds its annual meeting so they can first air their grievances with their peers. Or the faction may wait to initiate a split until it can secure foreign support, accumulate weapons, recruit more members, or gain the backing of influential local leaders. Either way, the precise timing of an organizational split comes down to a combination of strategy and preparation, and it is for this reason that studies focusing on when splits occur can produce misleading results. Perhaps what these studies reveal are the conditions most *opportune* for organizational splits and not the conditions that are most detrimental to organizational cohesion as they anticipate.[h]

Since existing research overwhelmingly focuses on Stage Three of splinter formation—the most visible aspect of an organizational split—it neglects the intragroup politics that occur in Stages One and Two. It overlooks how factions form, what motivates them, to whom they appeal, and how they might be avoided. In some very important ways, however, the dynamics taking place in these earlier stages are self-reproducing; that is, "the movement of initial events in a particular direction induces subsequent events that move the process in the same direction."[49] In this case, although a subgroup forms early on, *why* it forms and *who* joins will critically shape the subgroup and its parent organization well into the future, affecting the behavior and internal dynamics of both. In the next section, I show how two components of this process are particularly influential and result in very different trajectories for emerging splinters.

Variation in Organizational Fractures and Splinter Behavior

While factions commonly form in and then break away from armed groups, one feature that differentiates them is their motivations: in other words, the faction's shared dissatisfaction and desired change that is its purpose. This is what injects variation into this relatively structured process, and it has important implications for the trajectories of breakaway groups.

---

[h] As for how long this process from faction formation to faction exit takes, there is no clear answer. Some groups can remain at Stage One, with relative internal unity, or even at Stage Two, when relatively cohesive factions are operating within an organization. Sometimes a parent group's leaders successfully forestall a contentious split during Stage One or Stage Two, or maybe a harmonizing event occurs. Owing to the uncertainty surrounding the timing of the split itself, there is no way to anticipate how long this process will typically endure.

The first dimension that shapes the outcome of organizational ruptures has to do with the specific disagreements involved. In other words, *what type* of grievances motivate factions to form and break with their parent? These disagreements are important because they influence the types of individuals that are drawn away into breakaway groups—militant hardliners, ideological purists, personal followers, and so on—and what preferences they hold for their organizational future. It also tells us about the type of members from which the parent group will be relieved. Overall, I find that splits within armed groups occur for a variety of reasons, but they can generally be classified as occurring over three issues: (1) leadership, (2) strategy, and (3) ideology. Splits involving strategic disagreements tend to drive members with hardline preferences into breakaway groups, and this type of split is associated with a splinter's eventual strategic and tactical escalation.

The second dimension concerns *how many* grievances are motivating the subgroup to break away. This has important effects on whether the preferences of group members are aligned or not. The initial level of preference alignment matters because as grievances increase, the splinter appeals to a more diverse pool of members that hold distinct visions for their organizational future. As subgroups become less homogeneous, they become more susceptible to internal tensions and external manipulation that undermine their ability to survive.

Taken together, the implication is that variation in organizational splintering is due to variation in why factions form. When it is hardliners breaking away over strategic disagreements, radicalization should follow. When a diverse group of fighters breaks away over a multitude of disagreements, survival becomes increasingly difficult. Only by understanding the contours of faction and therefore splinter membership can we make accurate predictions about how they will behave.

Types of Internal Disagreements

It is important to understand the disagreements that cause factions to form and that cause armed groups to break apart. We know that subgroups attract like-minded individuals who coalesce around shared ideas, dissatisfaction, and preferences for their organizational future. This is because individuals rationally navigate subgroups and join those to which they feel a specific affinity.[50] We also know, quite obviously, that membership in a faction *before* a split predicts membership in the resulting breakaway group *after* a split.[51] So if we can understand the disagreements within armed groups, we can plausibly predict important aspects of breakaway groups once they become independent.

When it comes to armed groups, three issue areas explain the majority of disagreements. They are disagreements over strategy, ideology, and leadership.[i]

First, armed groups sometimes rupture over strategic disagreements that stem from tactical or strategic choices. Tactical disagreements have to do with specific operational decisions involving the types of attacks to launch and the types of targets to hit. Strategic disagreements, on the other, are more about the broader plan of action to achieve the group's goals. They could include differences regarding the use of violence in general, the utility of attacking civilians, or the value of accepting a ceasefire or negotiation. They might also involve differences over how and when to pursue their goals.[j] Taken together, these disagreements concern the utility and application of force in relation to group objectives.

Strategic disputes most commonly occur over the decision to escalate or continue the use of violence, with breakaway factions holding preferences for more extremist behavior. While there are, notable cases of splinter organizations breaking away to become more moderate, as Ansaru initially did with Boko Haram, this, is relatively rare. Empirically, notable cases of militant fragmentation demonstrate this trend, with the overwhelming majority of breakaway groups in Northern Ireland, Syria, Iraq, Spain, and Israel/Palestine conforming to this pattern.[52] Theoretically, this is because more moderate dissenters have many options and do not need to create new groups to represent their views. As Bueno de Mesqiuta writes, "moderate splinters are uncommon [because] the more moderate end of the ideological spectrum is already dense with political organizations."[53] Instead of breaking away to form a more moderate armed group, those seeking moderation will commonly defect to other armed groups, to political parties, or they will simply abandon violence altogether. For instance, when peace negotiations were taking place between Israel and the Popular Front for the Liberation of Palestine—General Command (PFLP-GC) in 1999, its leader, Ahmad Jibril,

---

[i] Of course, armed groups might experience other types of disagreements, but they do not necessarily suffice to rupture the organization.

[j] But not the value of those goals, or whether to embrace more goals.

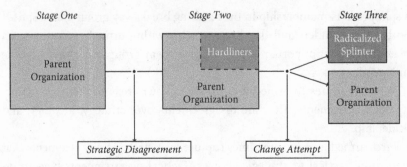

Fig. 2.4 The process of a strategic rupture

took a hardline stance and refused anything short of major concessions. More moderate leaders who opposed this hardline stance did not break away and form their own group, but they instead quit and renounced violence.[54] This pattern of moderate defection, which is central to research on ethnic defection, has repeated itself numerous times.[55]

Disagreements over the escalation or continued use of violence catalyze subgroups and eventually splinters that appeal to a specific subset of members. Rhetoric about increasing the use of force will attract fighters from the parent group who are hardline, extremist, and dissatisfied with their parent's tactical and strategic choices. These individuals want to increase violence. Once these subgroups become independent, they can also attract new, like-minded members from other organizations or even the general population who feel the same way. For instance, the Irish National Liberation Army broke away from the Official IRA in 1974 when they agreed to a ceasefire with the British government. After becoming independent, their vocal criticism attracted similarly disaffected hardliners from other groups like the Provisional IRA.[k]

Second, armed groups also break apart over ideological disagreements. I take a broad view of ideology as encompassing group identity, goals, religion, and beliefs. Ideological disagreements are often instigated when group leaders adopt new ideologies, whether they are religious, economic, or social in nature. For instance, ideological disagreements ravaged european armed groups in the 1960s and 1970s when Marxism was gaining popularity, generating disputes over its relative merits and its incorporation into group

---

[k] This example again underscores how factionalism does not always precipitate organizational splits. In this case, factions opted to join the new INLA instead of breaking off on their own.

doctrine. Yet, ideological disagreements often erupt over commitments to existing goals and ideologies as well, with factions claiming that group leaders lack sufficient dedication to core beliefs. For example, one of the reasons that Al-Shabaab split from the Islamic Courts Union in 2004 was the Courts' purportedly waning commitment to Sharia law.[56]

Subgroups that form, and splinters that eventually emerge, from ideological differences attract members who sympathize with their ideological cause. These ideological preferences can be based on a variety of issues, ranging from specific disagreements over post-conflict aspirations to economic principles. Many ideological disputes tend to reflect differences of degree rather than differences in fundamental beliefs. It would be rare, to say the least, for members of a Jihadist militant group to break away and form a splinter devoted to Shiite principles. Rather, most take place over the incorporation of new objectives or ideologies or, conversely, over degrees of contention surrounding existing objectives or beliefs.

Third, armed groups can break apart over leadership disputes. These disputes are not uncommon because individuals face strong incentives to be in command. Such a position confers prestige, respect, and sometimes a significant amount of money. The Islamic State was famously bringing in nearly $3 million in revenue per day at its height. In terms of individual leaders, Bosco Ntaganda, who led the group known as M23 in the Democratic Republic of Congo, regularly brokered illegal gold sales worth tens of millions of dollars.[57] Consequently, there is no shortage of individuals seeking to climb the leadership ladder and wrest control for themselves. Even if money is not involved, personal disputes can quickly spill over into organizational politics. For instance, a rift occurred within Jaysh-i-Muhammad in Pakistan after its leader, Masood Azhar, expelled a dozen members for conducting unsanctioned attacks. This prompted other members to break away from the organization to form a new group called Jamaat-ul-Furqan.[58]

Organizational splits caused by leadership disputes are relatively easy to identify. In stark contrast to many of the examples discussed so far, they do not involve the group's organizational trajectory. These ruptures do not involve disagreements over divergent strategic plans or ideological interpretations, but they stem from competing preferences over who should be in charge and whether one person is sufficiently respected or listened to. Here, one must recall that despite the violent aims of these organizations, armed groups are still compromised of men and women who face many of the same challenges and interpersonal disputes as other groups. One of the challenges for members of *any* organization is simply getting along.

When leadership disputes polarize a group, subgroups and eventually splinters form around particular individuals. When the Rally for Congolese Democracy held an internal election to determine their new leaders, for instance, some of the losing candidates broke away with their most loyal followers.[59] In this way, the cohesion of breakaway groups motivated by leadership disputes is typically weak: rather than breaking from their parent over differences in strategy, tactics, or ideology—all of which frame, contextualize, and align actions and beliefs of group members—these individuals are instead motivated by personal allegiances.[60] Different preferences for tactics and post-conflict goals might therefore lie under the surface and come into tension before too long. And, if the group is primarily bound together in a sort of cult of personality, then the future may be bleak if that individual dies or is assassinated. While this possibility was raised after the United States killed Osama Bin Laden, it overlooked Al-Qaeda's significant decentralization that had taken place in recent years.[61]

As this discussion suggests, it is imperative to understand *why* organizational ruptures occur since the motivations underlying schisms shape who stays behind, who departs, and what those departing seek to accomplish. Not every split is driven by the same type of faction, the same type of disagreements, or the same types of individuals. We also cannot deduce this information solely from what is occurring around or to the group. Understanding intragroup politics are essential.

A Preference for Radicalization

To begin with, it is important to note that armed organizations do not adopt tactics and strategies haphazardly. They are rational actors—albeit with some cognitive, informational, and other limitations—and the choices they make conform to internal cost-benefit calculations. For example, consider how the Tamil Tigers (LTTE) adopted suicide bombings.[62] This decision was made by Velupillai Prabhakaran, the leader of the LTTE, after witnessing how Hezbollah's 1983 suicide attack against an American Marine barracks in Beirut hastened the US's withdrawal from Lebanon. As Prabhakaran explained, "With perseverance and sacrifice, Tamil Eelam can be achieved in 100 years. But if we conduct Black Tiger [suicide] operations, we can shorten [it]."[63]

This shows how rational calculations and individual preferences shape the tactics and strategies of armed groups.[1]

Armed organizations are also, to varying extents, deliberative bodies. It is not only the opinion of a single leader that matters. As Crenshaw notes, "Acts of terrorism are committed by groups who reach collective decisions based on commonly held beliefs."[64] The composition of armed groups therefore strongly shapes the decisions they make. Some argue that the Kurdistan Workers Party (PKK) made a decision to embrace a more moderate approach after 1999 when one of its hardline founding members, Abdullah Ocalan, was arrested by the Turkish government. His removal fundamentally shifted the group's composition and resulted in clearly observable changes to their outward behavior.[65] Similar effects also occur when states employ conciliatory approaches that draw moderates out of armed groups, reshaping their internal composition. This leaves hardliners in control, and more violent operations usually follow.[66]

But is it even possible to distinguish between hardliners and moderates when talking about members of an armed organization? While armed group members are among the most hardline individuals within the overall population, research makes it clear that there is still significant variation among those who become a terrorist or insurgent. For example, a Congressional Research Service report made this point when it explained that "A psychologically sophisticated policy of promoting divisions between political and military [terrorist] leaders ... is likely to be more effective than a simple military strategy based on the assumption that all members and leaders of the group are hardliners."[67] Indeed, despite engaging in violence, some members may hold more moderate views on civilian targeting and the value of political compromise.[68]

With this in mind, breakaway groups that contain greater numbers of tactical and strategic hardliners with preferences for extremist violence are expected to radicalize. That is, when the distribution of preferences within an organization favors radicalization, escalation becomes likely. This happens for two specific reasons.

First, and most obviously, tactical and strategic hardliners will *directly* influence the discourse and decision-making of breakaway groups, especially

---

[1] Without Prabhakaran's explicit desire to bring suicide bombs to Sri Lanka, it is uncertain whether the tactic would have arrived on its own. Of course, Horowitz (2010) notes that "sometimes desire is not enough to adopt an innovation." While a desire to introduce a tactical innovation is perhaps not a sufficient condition for tactical success, it does seem like a necessary one.

when they hold positions of power. When potential plans are discussed, these individuals will lobby for and agree with more radical approaches. As deliberative bodies, actions will flow logically from the majority's preferences, and it will be hard to prevent escalation when hardliners hold the majority.

Second, greater numbers of hardliners with preferences for radicalization will *indirectly* push splinter groups toward more violent operational profiles. This is because the leaders of armed groups often enact policies to satisfy particular subgroups to maintain unity and ultimately, to survive. In his model of terrorist fragmentation, Morrison notes that:

> In relation to terrorist organisations it is observed that in order to avoid the departure of significant sub-groups of an organisation a formerly moderate membership may at times have to radicalise their tactics and strategies. Therefore at certain stages the terrorist actions may aid in the survival of this organisation. This is a complex role for the organisation to play as if they over radicalise they risk losing the more moderate membership and external support. However, if they are not radical enough the risk lies in losing the more radical elements of the membership and support.[69]

In other words, when armed groups possess meaningful numbers of hardliners, their leaders—even if they are not hardliners themselves—may ramp up violence for the sake of organizational cohesion. Since hardliners usually represent a particularly dangerous, capable subset of the group, leaders may fear internal conflict, coups, or a split if their desires are not met. This has been the case with the Palestinian group, Hamas: leaders are forced to take hardline negotiation positions over fears that "hardliners would leave Hamas and join Al-Qaeda-inspired salafi-jihadi groups in Gaza that remain ideologically committed to violence in the name of religion."[70]

Ultimately, when there is a large subset of hardliners within an armed group, their policies and actions will tend to be more extreme, leading to an observable increase in levels of radicalization. And the organizational fractures most likely to produce splinters dominated by hardliners are those are those taking place over strategic disagreements. These splits overwhelmingly occur when a dissatisfied faction seeks to renew or expand the use of force, pulling in hardline militants who have an appetite for violence, and driving their new group towards tactical and strategic extremes.

The Number of Internal Disagreements

In addition to the types of disputes shaping patterns of radicalization, the number of disagreements between a splinter group and its parent affects whether newly formed breakaway groups survive. In the previous section, I show how different types of disagreements attract different types of members—with splits over strategy commonly appealing to hardliners. However, it is not always the case that only a single grievance is responsible for an organizational fracture. When splits take place over multiple grievances, the new organizations can attract a heterogeneous mix of recruits—hardliners, ideological purists, or simply those influenced by personal connections. When an organization is comprised of individuals with competing preferences for their organizational future, its survival is threatened.

To this end, organizational ruptures can occur either unidimensionally or multidimensionally. Unidimensional splits occur when a single, shared grievance motivates a faction to depart. Multidimensional splits occur when factions are motivated by several different grievances. Some examples may help demonstrate these differences. A unidimensional split divided Sendero Luminoso (the Shining Path) in Peru. Their captured leader accepted a peace deal with the government in 1994, and "This compromise caused a fracture in the movement, dividing it into two groups: those who wished to continue to carry out the vision of Sendero through violence, and those who wished to surrender."[71] This organizational split stemmed from a singular disagreement that polarized the group into two competing factions. On the other hand, a multidimensional split divided the Popular Front for the Liberation of Palestine when members grew "uncomfortable with the organization's Marxism-Leninism as well as with some of its tactical decisions."[72] The Moro Islamic Liberation Front's departure from the Moro National Liberation Front in 1977 was similarly multidimensional: it stemmed from intense leadership and ideological disputes. Members were frustrated with Nur Misuari's "dictatorial and corrupt" style, and they also endeavored to create an Islamic state, in contrast to the secular state desired by Misuari.[73]

The number of grievances in an organizational split matters because it shapes a group's *organizational niche* and therefore the cohesion and preference-alignment of its members. Developed in the fields of organizational theory and ecology (and before that, environmental ecology), a group's organizational niche is essentially who it appeals to and extracts resources from;[74] it "represents the position or function of an entity, such as an organization or population of organizations, within a larger community environment."[75] Whether a political party, religious sect, business firm, or even armed group, all organizations occupy some particular niche that may be distinct from or overlap with the niche of others. Niches have been measured in many ways: according to the age and location of students enrolled in a given school,[76] the size of engines produced by car manufacturers,[77] and the types of beer produced by breweries.[78]

There are rationales for groups to both expand and contract their niches. On the one hand, expansion provides access to a larger potential population of supporters. A political party, for instance, claiming to represent everyone from environmentalists to the coal industry can theoretically broaden its reach to many voters. This is why most presidential candidates in the United States cannot stray too far to the left or to the right, but instead remain relatively centrist to maintain somewhat broad appeal.[79] But there is a trade-off to expanding one's niche that leaves groups with two uninspiring choices:[m] follow an agenda that minimally appeals to the broad coalition of supporters, or follow an agenda that greatly appeals to only part of the coalition.[80] Obviously, both of these options are flawed: the first leaves every member somewhat dissatisfied, while the second leaves many members mostly dissatisfied.

On the other hand, smaller niches will tend to contract a group's potential pool of supporters. This should make it easier to design an agenda that is broadly and enthusiastically appealing, sparing groups from the dilemma identified earlier, but it also limits their maximum amount of support. Consider some of the smaller political parties that form in parliamentarian systems, like the Pirate Party in Sweden. Initially formed with a focus on copyright and patent laws—a small niche—they only managed to earn about 35,000 votes (0.63 percent) in their first election in 2006. After expanding their niche to include "access to free communication, culture and knowledge" and a focus on privacy issues more generally, the party managed to gain nearly 335,000 votes (or 7.13 percent) in subsequent elections. Owing to these consequential effects, research links niche size to almost every aspect of a group's behavior including the odds of survival,[81] internal competition

---

[m] Although this dilemma plagues organizations of all kinds it is especially severe for militant organizations that often critically need more members but also depend on internal cohesion and effective control to survive.

and cooperation,[82] performance (or "fitness"),[83] and ability to generate resources.[84]

In line with existing research from other fields, we can conceptualize the organizational niche of a splinter as the grievances that motivate a faction to form and to exit its parent group.[85] These grievances are central to the faction's identity, to how it differs from its parent, and to whom the faction appeals from the parent organization. Combined, this makes it a good approximation of its niche. Multidimensional splits, where more than one disagreement is prompting the faction to depart, will generate large niches for the emerging splinter group. In these cases, the faction will draw support from a wider population who can see its interests represented by this new organization. Unlike splits that polarize the group into distinct camps around more concentrated opinions, however, factions from multidimensional splits can attract members with only slightly overlapping preferences. While this should help fledgling splinter groups appeal to a greater segment of the parent organization, that audience tends to be more diverse. Unidimensional splits, on the other hand, create groups with comparatively smaller organizational niches. What they hope to accomplish and how they hope to change is more narrow and focused owing to the singular grievance with their parent. In economics terms, these members should have more concentrated preference profiles;[86] that is, they will tend to hold more aligned preferences for political goals, use of violence, ideological interpretation, and so on.

To illustrate, consider a faction that forms within a rebel group because of a desire to escalate violence or to reject a peace deal (e.g. Figure 2.5). That faction will draw in members who seek to escalate, expand, or renew their operational tempo. This was the case when several factions broke off from the Sudan Liberation Movement over their objection to the Darfur Peace Agreement.[87] Yet, a faction that forms to escalate violence but also to embrace Marxist ideology (e.g. Figure 2.6) can draw in disaffected members who are drawn to both ideals, but also perhaps to just one or the other. Some revolutionary organizations in Italy dealt with this exact problem, ending up with an overly broad coalition that undermined the group's cohesion. In effect, "the encounter of the political entrepreneurs with the alienated youth produced short explosions of violence: the preferences and attitudes of the two groups were too heterogeneous, and a stable alliance was therefore impossible."[88] Indeed, the members of an armed group will not necessarily value all of their group's goals

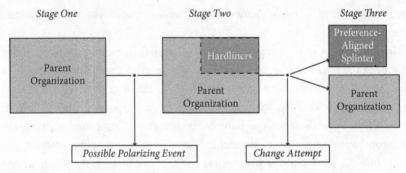

Fig. 2.5 The process of a unidimensional (strategic) rupture

equally.[89] But by narrowing their goals, and in effect their organizational niche, armed groups can limit their appeal to more like-minded fighters. This reduces uncertainty about what they stand for and attracts a more preference-aligned membership base. In the following section, I explain how this has significant ramifications for their survival.[n]

Competing Preferences and Survival

How does niche size affect a splinter group's odds of survival?[90]

First, assuming that multidimensional grievances appeal to members with heterogeneous preferences, then splinter groups emerging over multiple disagreements will experience greater rates of disagreement, infighting, defection, and leaking information to authorities.[o] Summarizing this situation, Milliken and Martins write that "the greater the amount of diversity in a group ... the less integrated the group is likely to be and the higher the level of dissatisfaction and turnover."[91] Dissatisfaction occurs because groups with larger niches find it difficult to fully satisfy their members who hold diverse goals. Simply put, "The larger the niche size, the more difficult it is for the organization to serve any particular part of the niche as well as a

---

[n] One may wonder whether it is possible that a multidimensional fracture could in fact *narrow* the preferences of groups members and produce greater homogeneity (in stark contrast to what I suggest) by appealing to members who disagree with the status quo across multiple dimensions. However, I am dubious of this explanation for two reasons. First, members of armed groups often have very different preferences, and this explanation supposes that there is a preexisting subset of members that could be brought together to form a splinter. This is unlikely. Second, when members disagree with their parent across multiple dimensions, it is rarely the case that all of these individuals value every disagreement. Some may be motivated to join by the strategic dispute, others by ideological disagreements, and others still may be following the faction's leader. As before, this makes a unified, multidimensional splinter unlikely.

[o] In a similar way, Pischedda (2018) finds that "ideological differences, disagreements over strategy, [and] different priorities" have led to conflicts *between* rebel groups as well.

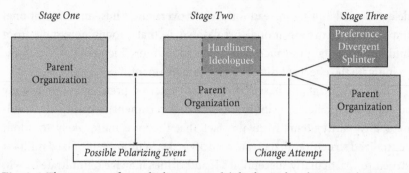

Fig. 2.6 The process of a multidimensional (ideological and strategic) rupture

more specialized organization could."[92] Over time, this dissatisfaction can motivate deviant behavior like leaking and defection as fighters reevaluate their allegiances and grow susceptible to government payoffs and coercion. Cooperation with and obedience to a splinter group are most likely to occur when individuals derive nonpecuniary rewards from doing so; in other words, when they "derive utility from working for and associating with the group."[93] Such utility is unlikely amid internal turmoil.[P]

Second, in situations where members of a subgroup have a greater diversity of preferences, leaders opt for hierarchical organizational structures that are better suited to micromanagement and control. This response to principal-agent problems among armed organizations is well documented.[94] Yet, principal-agent concerns should be less likely to emerge when groups experience internal consensus. As described earlier, consensus typically fosters cooperation,[95] allowing leaders to delegate responsibilities with fewer fears of insubordination or unsanctioned behavior. In splinter groups with heterogeneous membership bases, however, leaders may be hesitant to

---

[P] Although these nonpecuniary awards are difficult to identify, one example is how ideological groups often "[rely] on the nonpecuniary rewards of 'fighting the good fight'" (Scott Gates (2002). "Recruitment and Allegiance: The Microfoundations of Rebellion." *Journal of Conflict Resolution* 46.1, pp. 111–130, p. 114) to motivate their members. In other words, nonpecuniary benefits refer to the personal, immaterial, and intangible benefits that individuals derive from contributing to an organization's broader objectives. Nonpecuniary benefits motivate individuals to support the organization's strategic objectives absent positive (material rewards) or negative (punishment) inducements by the leadership. Although all organizations tend to be motivated by a mixture of rewards and punishments, armed groups with more internally aligned preferences can more easily and more consistently draw upon nonpecuniary awards to motivate their members and keep them in line. In this regard, shared preferences are helpful because "For agents to derive solidarity benefits or functional benefits, they must derive utility from working for and associating with the group" (ibid., p. 115). If individuals are often disagreeing with and arguing against other members, there is little chance that nonpecuniary rewards will produce cooperation. Preference alignment can therefore be critical when it comes to motivating cooperation among members within an armed group.

delegate at all, opting instead for greater oversight. Thus, an unintentional implication of organizational breakdown is that groups emerging from multidimensional grievances will often adopt more hierarchical structures to mitigate the effects of diversity and disagreement.[96]

Third, organizations beset by divergent internal preferences are acutely susceptible to defection, infiltration, and government targeting. This vulnerability partly stems from the fact that they are more likely to adopt centralized structures that create security risks. Why are centralized militant structures particularly susceptible? If authorities manage to infiltrate or win over a defector from a highly centralized organization, the information the government obtains will likely be more damaging to the group as a whole. As Enders and Su note, "Since communication links to the leadership are all direct," in hierarchical structures, "[it] is not especially secure, since every node has the possibility of providing useful information about the location of the leadership."[97] On the other hand, decentralized, cellular networks are more effective at blunting the effects of infiltration and defection as they compartmentalize both their operations and their units on the ground. Under these conditions, "Efforts at sowing discord and dissension within terrorist groups will probably be less effective on cells which are only very loosely connected with other cells and leadership through one member or another."[98] Gaining information from or access to a single unit that is part of a decentralized group does not necessarily threaten the entire organization.

Fourth, and finally, the sum effect is that unidimensional and multidimensional motivations shape the ability of groups to survive.[q] This idea finds support in research on organizational ecology where specialist firms—those with smaller organizational niches—tend to outperform and survive longer than generalist firms.[99] For armed groups operating in a threatening, anarchic environment,[100] however, infighting, defection, and leaking can quickly precipitate decline. And, while adopting a hierarchical structure may seem like a solution, it entails a cost: structures that prioritize internal control do so at the expense of security. As Shapiro notes:

---

[q] By "survive" I am not only referring to the organization's defeat at the hands of the state, but also to demise via fratricide, further splintering, and violent competition for control.

When the preferences of leaders and agents are not completely aligned, the covert nature of terrorist groups necessarily implies agents can take advantage of the situation to act as a preferred, rather than as their principles would like.... The costs for terrorist groups are obvious; monitoring agent activities requires additional communications and record-keeping, which thereby increases the risk of death or imprisonment for everyone in the group.[101]

As a result, a splinter groups solution to its internal disagreements may ultimately hasten its downfall.[r]

The Rupture of ETA

How well does this model of organizational splintering and its expectations conform to historical events? Here, I demonstrate its veracity with a brief case study of a rupture experienced by the Basque terrorist group Euskadi Ta Askatasuna (ETA). While the group experienced many splits, I focus on one of the most significant which produced the deadly and durable ETA-*militar* in 1974. This split occurred after a series of setbacks prompted group members to rethink the status quo. A faction soon coalesced around common grievances and ideas for the future, and upon recognizing the limits of their influence, these members broke away to realize their vision. This process is depicted in Figure 2.7.

---

[r] While any armed group with members holding heterogeneous preferences is susceptible to internal and external threats, overcoming these threats will be especially challenging for newly formed splinters. While they experience the typical "liability of newness" (Josef Brüderl, Peter Preisendörfer, and Rolf Ziegler (1992). "Survival Chances of Newly Founded Business Organizations." *American Sociological Review*, pp. 227–242)—challenges of recruitment, resource acquisition, and a lack of credibility—they must also contend with potentially hostile parent organizations and a public that may be reluctant to support yet another group. This helps to explain why heterogeneity can be severely problematic for new groups—especially splinters—but not necessarily for established organizations. In addition, the "sunk costs" of membership in new organizations is relatively low (Donatella Della Porta (1995). "Left-Wing Terrorism in Italy." In *Terrorism in Context*, pp. 105–159) insofar as members have not yet developed interpersonal bonds or committed many violent acts that might prevent exit, defection, and leaking. Indeed, in related fields, research finds that niche concerns and positioning vis-à-vis established groups are particularly relevant for "newcomers" (Josephine Andrews and Jeanette Money (2008). "Champions and Challengers: Ideology and the Success of Political Parties in Established Party Systems.") Ultimately, these challenges may prove insurmountable for splinter groups and play a role in their demise.

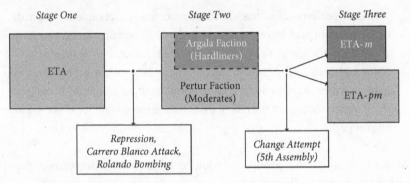

Fig. 2.7 The rupture of ETA in 1974

Stage One: Organizational Unity

ETA is a terrorist organization founded in the Basque region of northern Spain around 1950. The Basque region (El País Vasco), or the Comunidad Autónoma Vasca, possesses a distinct culture and ethnicity. The native language, Basque, is separate from the Castilian spoken elsewhere in Spain. The Basque region also has a long history of relative autonomy. Following the combined brutality of the Spanish Civil War and the dictatorship of General Francisco Franco, however, citizens of the Basque region felt a renewed urgency to manage their own affairs independent of the central government in Madrid. It is ultimately this goal—territorial independence—that motivated terrorist organizations like ETA to form.

While ETA's main goal is Basque autonomy, it has also been motivated by various ideologies like Marxism and communism. These ideologies were generally subordinated to ETA's primary aim of territorial independence, but they did sometimes influence the group's strategic thinking in meaningful ways. For instance, at various ETA Assemblies (major organizational meetings) the group debated the merits of their violent strategy and how it might function alongside or eventually lead to a proletariat uprising, the bedrock of any Marxist struggle. Specifically, ETA members considered how to eventually transition from a sporadic, terrorist-style campaign to a *lucha de masas*—mass struggle. These sorts of debates were not uncommon among European terrorist organizations in the 1960s and 1970s.

Owing to these intense ideological and strategic debates, ETA experienced many organizational ruptures. At various times, subgroups split off, merged, defected, or formed political parties. These ruptures were significant: in line with my expectations outlined above, these splits fundamentally shaped and reshaped the organization's membership over time by separating out dissatisfied factions with competing views. According to Clark, "Each time [one] of these groups left ETA, those remaining tended more and more toward homogeneity of thought and purpose... as a result of these debates, splits, and consolidations, ETA emerged each time radicalized, more intransigent and more deeply committed to armed struggle."[102] Accordingly, the strategic and ideological evolution of ETA is also a story of organizational fractures.

While ruptures were relatively common, one of the most significant took place in 1974. This produced two distinct organizations: the bulk of the group's members stayed with the parent organization, henceforth going by the name of ETA-*político-militar*; and a minority of members broke off to form ETA-*militar*.

Polarizing Events

Three polarizing events are responsible for ETA's mounting internal tensions in the late 1960s and early 1970s. The first was a brutal crackdown by the Franco regime that the group barely survived; the second was ETA's assassination of Luis Carrero Blanco in 1973; and the third was the bombing of Madrid's Cafe Rolando in 1974.

The first, and most enduring, event that spurred debate within ETA was the Spanish regime's brutal campaign of repression in response to the group's first major violent attack in 1968. While ETA had existed for years prior, it was primarily a forum for debate, nonviolent action, and planning for future resistance. Of course, violence was not entirely absent, but such actions were more sporadic and often flowed from protests that ended with violent clashes between Basque youths and the police.

This changed with ETA's 5th Assembly meeting in 1968, which ended up being a watershed moment in the group's strategic evolution. It is at this point that ETA decided to fully embrace violence. This decision was prompted by the death of ETA's popular leader, Txabi Etxebarrieta, who was killed in a gunfight with police. As testament to the respect Etxebarrieta commanded, cities and town throughout the Basque country held funerals for him even though such events were outlawed by the national government.[103] Ultimately, this was the spark that turned ETA's debates about violence into reality. Basques were fed up with Franco, who had suppressed expressions of their ethnic and cultural identity for years, and Etxebarrieta's death was the final straw.

Not long afterwards, ETA initiated its first planned attack. The group targeted a brutal police chief in the city of San Sebastián named Melitón Manzanas. He was notorious for torturing suspects and opposing Basque nationalism. In August of 1968, ETA members ambushed and killed him at his home. But ETA would pay a high price for their action. Soon after learning of Manzanas' death, Franco launched a swift crackdown and placed the entire Basque region under siege from 1968 to 1972. To illustrate the severity, in August 1968 there were reportedly six hundred arrests throughout the Basque region; in 1969, there were over 2,000. These arrests not only decimated ETA's leadership, but they targeted almost anyone associated with the Basque independence movement.[104] While ETA expected Franco to retaliate, it underestimated the scale of his response. This led to serious disillusionment and disagreement within the organization over their decision to embrace violence at the 5th Assembly.[105]

The second polarizing event came after the group emerged from Franco's brutal crackdown and recommitted itself to armed struggle. Although it would have been impressive for ETA to simply survive this period, it instead planned some of its most daring operations. The most well known was the 1973 assassination of Franco's prime minister and second-in-command, Luis Carrero Blanco. A small ETA cell in Madrid uncovered that Blanco was vulnerable on his way to and from Catholic Mass each Sunday. They monitored him for over a year and developed a plan to tunnel under the street that Blanco took each week. They posed as sculptors working on an art project and hid rubble from their tunneling operation in plastic bags. Once the site was prepared, the team placed nearly 80 kilograms of explosives directly below Blanco's route. On Sunday, December 18, 1973, the operation succeeded: the explosives were detonated as Blanco drove overhead. His car was catapulted more than five stories into the air, killing him in the process. The assassination left Spanish politics in limbo, a precarious situation that was exacerbated when Franco's health sharply deteriorated only a few months later.[106]

Carerro Blanco's assassination—dubbed by some as the "greatest mistake in the history of ETA"—was polarizing because of the political uncertainty it generated. Would this create an opportunity for Spanish democracy to flourish? Did the armed campaign need to continue now that the Francoist regime had seemingly been decapitated? These two considerations prompted ETA members to contemplate whether their objectives could now be achieved through politics rather than violence. Unsurprisingly, these developments inflamed the subset of members of who had already grown weary of the armed campaign after the brutal repression of the late 1960s and early 1970s. As a result, some who favored the predominantly political path to Basque independence—rather than one combing politics *and* violence—started defecting from the organization. Many of them joined groups like Basque Socialist Party, the Popular Socialist Party in France, and others.[107] The aftermath of Blanco's killing demonstrates how moderate members of an armed group often defect from a group rather than departing as more moderate splinter groups. It also underscored how ETA was changing now that its most politically minded members were departing, leaving a younger generation of hardliners in control.

The third and final event that generated tension within ETA was the attack colloquially known as the Rolando Bombing, named for the Cafe Rolando in Madrid where it occurred.[s] On September 13, 1974, a bomb went off in the cafe, killing 13 and injuring over 50. The cafe was targeted because it was located across the street from the Spanish government's Bureau of Security. Those responsible for the attack hoped to kill members of the Bureau who might be getting lunch and to send a message that no location is safe. In many ways, the Rolando Bombing was an aberration from the typical ETA operation: the majority were highly discriminate, targeting and killing only a few people each time. Between 1968 and 1975, for instance, ETA attacks killed a total of 45 people—13 of whom died in the Rolando Bombing. The operation was therefore highly divisive, inflaming debates between moderates and hardliners over the relative value of predominantly political and predominantly military approaches.[108]

By the early to mid-1970s, ETA faced severe internal tensions. The group was unsettled over its strategic orientation, with some favoring a violent, hardline approach and others hoping to moderate violence in favor of a political solution. Notably, these debates did not stem from a single government action or from attributes of the group's leadership. Instead, they reflected a combination of state repression, disagreements about specific operations, the potential for a political opening, and a lack of tangible success. Against this backdrop, members began to coalesce into distinct factions with competing visions for the group's organizational future.

---

[s] While ETA initially denied responsibility for the attack, twenty days after it took place, the group officially acknowledged their complicity in 2018.

Stage Two: Faction Formation

Two distinct factions soon emerged over the issue of strategy, approximating a unidimensional strategic split. The first was led by José Miguel Beñarán Ordeñana. Also known as "Argala," he was the head of ETA's *Frente Militar*, the sub-organization responsible for the group's most high-profile attacks including the assassination of Carrero Blanco. What tied the Argala faction together was its members' firm belief in the primacy of the armed campaign to achieve Basque independence. They aimed to become "the 'revolutionary vanguard' dedicated to armed combat."[109]

The second faction was led by "Pertur," Eduardo Moreno Bergareche. Despite his unsolved disappearance only a few years later in 1976, Pertur had an enduring effect. He was responsible for the group's simultaneous political *and* military strategy. He did not view the military option alone as sufficient to achieve Basque goals, and he saw value in a potential negotiated settlement. In effect, the Pertur faction tended to be much more moderate than the Argala faction.

While tensions had been building in ETA throughout the early 1970s, things escalated when the Frente Militar ramped up its military operations in 1974. Against this backdrop, ETA held one of its more significant meetings—called a Biltzar Txikia (or a "little assembly")—that convened the group's most important members. Pertur's faction controlled a clear majority, giving him the votes to pass a significant internal reorganization and strategic reorientation that moved ETA away from violence and toward more conventional politics. In particular, they approved amendments to abandon several important ETA subcommittees like the Frente Militar led by Argala. In its place, ETA's new committees would each take responsibility for simultaneous political, military, and propaganda actions. These changes angered the hardline Argala faction immensely, but with their inferior numbers, they were unable to do much about it.

Stage Three: Faction Exit

Feeling betrayed by these institutional changes, the faction around Argala (which happened to be much of the now-defunct Frente Militar) issued a statement a few days later announcing its departure. While retaining the name ETA, they would henceforth be known as ETA-*militar*, whereas those remaining under Pertur would be ETA-*político-militar*. The reason for this split is nicely summarized by a passage from Soldevilla:[110]

In 1974, strategic disputes, and not ideological, resulted in ETA V fragmenting once again. A minority of members, those that only believed in violence, created ETA*m*...the majority of the organization came to call itself ETA*pm*...Pertur's project, that was subverting some of the fundamental narratives of what came to be called the "Basque conflict"...provoked an irate response from the most nationalist and militaristic members of the group, the berezis [i.e. Frente Militar] (in particular) who accused the others of being heretics who were preparing for the "liquidation of the armed struggle."[t]

Underlying the schism between ETA's Pertur and Argala factions was a fundamental difference in strategy; it came down to "pura estratégia y táctica política."[111]

More information about ETA's rupture can be gleaned from their internal discussions. One edition of ETA's *Zutiks*, essentially a newsletter, offered the following explanation of the split to the Basque people:

It's been two years...since the Basque community, with great pain and confusion, saw ETA divided. The reasons behind the departure were differences with respect to the organization's structure, [and] the different views towards the organizational units responsible for various components of the popular struggle.... The constant failures of the Political-Military organization are palpable demonstrations of the impossibility of managing political action by an organization that simultaneously engages in armed conflict and mass struggle in the current phase of the revolution.[112]

One may question whether this rupture should be characterized as a strategic, unidimensional organizational split and not about leadership.

---

[t] Author's translation. The original is as follows: En 1974 las disputas estratégias, que no ideológicas, hicieron que ETA V se fragmentase de nuevo. Una minoria de sus activistas, los que únicamente confiaban en la violencia, crearon ETA*m*...la mayoría de la organización pasó a denominarse ETA*pm*.... El proyecto de Pertur, que subvertía algunos de los fundamentos de la narrativa de lo que luego se ha venido a llamar "conflicto vasco"...provocó la irración del sector más nacionalista y militarista del grupo, los berezis (especiales) quienes le acusaban de ser un hereje que estaba planteanda la "liquidación" de la "lucha armada."

While both Pertur and Argala were critical to these events, their disagreement had little to do with each other and much more to do with the armed campaign. Each championed a fundamentally different point of view and their factions were organized around their respective strategies. Issues about control and leadership never featured into the debates, leaving strategy as the sole point of contention.

Ultimately, the effect of this unidimensional strategic split was to separate out the majority of hardliners from ETA into the new ETA-*militar*, leaving the moderates behind under ETA-*político-militar*. This flowed logically from the internal disagreements that preceded the rupture. As one ETA communication from the early 1970s succinctly notes, "*la actividad militar ha de ser asumida por una sola organización*." That is, military activity has to be directed by a single organization and not, as the rest of the document makes clear, one that also engages in politics.[113] The resulting organizational trajectories of both groups should therefore come as no surprise.

On the one hand, ETA-*pm* would only last for a few more years. They entered into numerous ceasefires with the Spanish state, successfully negotiated for prisoners to be released from Spanish jails, and spawned several political parties and activist groups. For instance, in 1976, ETA-*pm* created a new political party known as Euskal Iraultzale Alderdie, or later on, Euzkadiko Ezkerra.[114] ETA-*pm* formally disbanded in 1982.

On the other hand, ETA-*m* would survive and challenge the security of the Spanish state for decades. It was a homogeneous group of hardliners committed to violence. Despite setbacks due to infiltration and arrests, ETA-*m* remained committed to its cause and to its strategy. The group eventually swelled to over 1,000 members, and between 1976 and 1982, ETA-*m* was responsible for over six hundred attacks.[115] It would not be until 2017—more than 40 years after it formed—that ETA-*m* would officially disband.

Observable Implications

My theory has two overarching implications for how splinter groups will behave.

First, unidimensional splits—like that which produced ETA-*m*—typically produce more durable groups. With smaller organizational niches, these groups appeal to more cohesive and homogeneous subsets of fighters who share preferences for their organizational future. Although conditions are difficult for any new armed group, these in particular will have a high probability of survival owing to their innate advantages. Of course, unidimensional splits can occur for several different reasons, including strategy, ideology, or leadership disputes. Disputes over leadership typically yield fewer of these advantages since breakaway groups are centered around specific figures and not necessarily a guiding ideological or strategic vision. These groups might also be particularly susceptible to leadership decapitation operations.

Second, splinters emerging from strategic disputes are most likely to radicalize. These breakaway groups are usually motivated by the desire to expand or renew attacks so they should appeal to hardline fighters with strong preferences for extremist violence (like the Frente Militar and the Argala faction). Their presence will directly and indirectly steer the group towards more violent operational profiles. Even when strategy is only one part of a multidimensional rupture, however, radicalization is possible. The presence of just some hardliners can shape a group's internal discourse and promote violence. The odds of radicalization among groups emerging over ideological and leadership disputes, however, are low.

In the following three chapters I demonstrate that these predictions are correct. In Chapter 3, with an in-depth case study of fragmentation among the Irish Republican Army; in Chapter 4, with new data on the breakdown of armed groups across the globe; and in Chapter 5, with a case study of the Islamic State breaking from Al-Qaeda. While the case studies allows me to test the mechanisms linking organizational breakdown to the subsequent behavior of splinter groups, the quantitative analyses show how these processes operate among different types of armed groups, throughout time and across space. Yet, the two case studies are quite different and serve distinct purposes. The Irish Republican Army is a critical case: here, fragmentation was frequent, varied, and historically important. For my theory to be useful, it should be able to explain how these events unfolded and why such different splinter groups emerged. Conversely, the Islamic State is not a critical case, but a deviant case, mainly because this is not a clear example of organizational splintering: the Islamic State and Al-Qaeda existed independently, joined ranks, then again went their separate ways. The extent to which the two organizations fused is also open to debate. Accordingly, my goal here is not to finely test the theory and all of its mechanisms, but to evaluate the utility of my theoretical framework in a more challenging environment.

While I am concerned with two broad aspects a splinter group's trajectory—namely, survival and radicalization—my theoretical framework can be used to understand other outcomes as well. Some of these can be seen with ETA. For instance, in terms of the likelihood that negotiations succeed or are even worth attempting, it would be useful to know why splits occur, which factions depart, and which remain. The Spanish state could have surmised that ETA-*pm* was well positioned to negotiate after its split that left moderates from the Pertur faction in control. Negotiations with ETA-*m*, and even including them in any multi-party agreement, would have been futile. As for the likelihood of inter-group conflict, with ETA-*pm* and ETA-*m* following such different strategies and appealing to such different types of recruits, the likelihood of both direct competition (e.g. armed attacks) and indirect competition (e.g. outbidding for recruits) was low. And with regards to counterterrorist strategy, a quick campaign to isolate, monitor, and potentially arrest ETA-*m* members before the group could establish structures and routines to evade state forces would have been advisable. With members predominantly hailing from ETA's *Frente Militar,* many of those within ETA-*m* had already committed serious offenses against the state. And now, lacking moderate voices, their new group would only become more dangerous with time, whereas pressure could be relaxed on ETA-*pm* to support negotiations.

3

Conflict in Northern Ireland

As I have described so far, armed groups commonly fragment, break down, and produce new groups from within their ranks. This process has become so common that is difficult to think of a single conflict in which fragmentation has not occurred. While ubiquitous, there is still significant variation in terms of the *quantity* and the *impact* of group fragmentation on various conflict dynamics. In some cases, one can point to one split in particular that indelibly shaped a conflict's trajectory. This was the case when the Basque terrorist group ETA split into ETA-*militar* and ETA-*político-militar* in 1974.[a] But in other cases, splintering is more frequent and many fractures are significant. This was the case in Northern Ireland. As Horgan notes:

> To suggest that splits have typified the development of Irish Republican militant groups is an understatement. Throughout history Irish Republicanism has continuously split and factionalized. These splits have not just shaped Irish Republicanism, they have led to some of the most significant and influential events in recent Irish history.[1]

Northern Ireland is therefore a crucial case for my theory. If it is going to be useful at all, it should help make sense of the dynamics here. And it does: as I demonstrate in the following pages, splinter groups exhibited neither uniform nor random trajectories. Instead, how they behaved and the extent to which they survived can be traced back to their earliest days as a subgroup within their respective parent organizations. The disagreements that motivated them to break away ultimately had a meaningful, path-dependent effect on their long-term evolution.

From the standpoint of research methodology, Northern Ireland is also practically useful. Organizational ruptures produced splinters that survived,

---

[a] Though there were others, this one stands alone as an especially defining event.

Divided Not Conquered: How Rebels Fracture and Splinters Behave. Evan Perkoski, Oxford University Press.
© Oxford University Press 2022. DOI: 10.1093/oso/9780197627075.003.0003

radicalized, and simply operated in very different ways; they "vary in their size, geographic location, strategies, ideologies, structures, and, not least, personality."[2] This variation offers an excellent opportunity to compare and contrast group trajectories, to isolate confounding variables, and to uncover why differences emerged. Additionally, there is a wealth of information available to researchers about Northern Ireland in general and the conflict more specifically. In this chapter, for instance, I draw upon a variety of primary and secondary documents including intra-group communications, interviews that fighters conducted with journalists, declassified government reports, and other materials that I uncovered during months of research in Belfast, Dublin, and London. Taken together, these sources provide a comprehensive understanding of the conflict.

While I cannot study every splinter group that emerged in Northern Ireland, I focus on the breakdown of two high-profile organizations and the formation of two distinct splinters. These are the Irish National Liberation Army (INLA) that emerged from the Official IRA (OIRA) in 1974, and the Real Irish Republican Army (RIRA) that emerged from the Provisional IRA (PIRA) in 1997. I chose them as part of a most-similar case study design.[3] According to this research strategy, one selects cases that share key similarities but exhibit different outcomes, allowing one to rule out these similarities as explanations for the outcomes. In this regard, the INLA and RIRA share some important characteristics. One can trace the emergence of both organizations to their parent groups de-escalating and moving away from violence.[b] Both groups were initially led by charismatic dissenters who shaped the doctrine and structure of their new organizations;[c] both groups initially sought to influence their parent organizations before breaking away;[d] both splinter groups created political and military wings;[e] and finally, both split from two of the most active, high-profile organizations of their respective eras,[f] with the RIRA splitting from the PIRA and the INLA from the OIRA.[4] Despite these important similarities, the INLA and the RIRA exhibit very different operational profiles and patterns of survival.

---

[b] This is important because de-escalation likely catalyzes hardliners to break away. Having this occur in only one case might introduce unique dynamics.

[c] Since strong, charismatic leaders could be associated with hierarchy, this is important to rule out.

[d] Those that do not attempt to influence parent organizations before breaking away are perhaps more likely to be motivated by personal gain, leading to lower cohesion.

[e] This could plausibly influence a group's membership.

[f] Splitting from weak, insignificant organizations might lead to different trajectories.

To briefly summarize, the INLA and the RIRA are two Republican militant organizations that share the common objective of Irish reunification. Although both groups were motivated to form by similar events—the cessation of violent activities by their respective parent organizations—the INLA formed multidimensionally over strategic and ideological disagreements. I would therefore expect the group to be violent, owing to the hardliners attracted by the strategic dispute, but also less durable owing to the heterogeneity of its members. As anticipated, the INLA's diffuse agenda ultimately attracted a wide range of militants, socialists, and politically focused recruits into the organization. Since these members held starkly different views of their organizational future, feuds were common, cohesion was minimal, and hierarchy was strictly enforced. They have subsequently been described as "ephemeral splinter group"[5] that quickly devolved into factional infighting.[g] Yet, the INLA also exhibited extreme violence. This can be explained by its vocal opposition to the Official IRA's ceasefire in 1972 that attracted hardline members from both the Official IRA and even the Provisional IRA into its ranks. It was these members who ultimately motivated and personally perpetrated some of the deadliest violence of the entire conflict. Taken together, this provides strong support for my theory of how splinters form and subsequently behave.

Like the INLA, the RIRA formed in opposition to a ceasefire. The RIRA's parent group, the Provisional IRA, announced a cessation of violence in 1998. But unlike the INLA, it formed unidimensionally and was motivated solely by issues of strategy. Its members simply wanted to resume the armed struggle. I would therefore expect this to yield a durable, highly violent organization owing to its cohesive base of hardline supporters. Once again, my theory is correct. The RIRA neither shared the INLA's political nor socialist orientations so it attracted a relatively homogeneous group of Republican hardliners who consistently and fully embraced violence. And, since their objectives were focused and their niche small, its member were highly cohesive. With feuding minimal, leaders were able to decentralize operational control. As a sign of how resilient this made them, they were described as an "enduring threat" as recently as 2015—nearly 17 years after they formed.[6] As before, the experience of the RIRA from its initial formation and throughout its independence strongly supports my theory.

---

[g] Within five years, the group had devolved into separate camps claiming legitimacy over the INLA name.

In the following pages I provide a fuller account of the INLA and the RIRA, delving deeper into their formation and evolution. I begin with an overview of the conflict in Northern Ireland before moving on to the armed groups themselves. The findings from this chapter not only corroborate my theory, but they underscore the logic of my broader theoretical framework. That is, splinter groups—and armed groups more generally—share many similarities with other collective entities, and by studying their membership and related group dynamics, we stand to learn much about them.

Background: The Historical Roots of Irish Republicanism

The conflict is Northern Ireland is largely about two separate but interrelated issues: first, the partition of the island of Ireland that occurred in 1921 and continues to this day; and second, the British presence in Northern Ireland. The partition of Ireland is significant as it ended the United Kingdom's direct control over the Irish island established in 1800. The British reigned over Ireland until 1921 and while some regional powers were devolved during this time, the island witnessed repeated armed rebellions to restore full independence. These rebellions occurred in 1803, 1848, 1867, and 1916. The last, termed the Easter Rising or the Easter Rebellion, began on Easter in 1916 and was the most successful one of all. The United Kingdom, however, quickly moved to contain the uprising, and its violent crackdown galvanized the Irish population and inflamed tensions even more. This unrest led to the Irish War of Independence in 1919 that raged for nearly three years and killed thousands on both sides.

A break in the fighting eventually occurred in 1921 when the British offered a peace treaty. Unsurprisingly, the offer was far from what the Irish wanted. The treaty stopped well short of granting full, unqualified independence. Instead, it set the groundwork for the eventual partition of Ireland—with 26 of Ireland's 32 counties forming a new Irish Free State and "qualified autonomy" for the other six that would remain part of the British Commonwealth.[7] The terms of the treaty polarized those fighting on behalf of Ireland. As Richard English notes, "The Treaty was not the republic, but it offered significant freedoms, and if it was rejected then how long would the IRA be able to hold out, if faced with intense war?"[8] Those who favored the treaty ultimately won out and the Anglo-Irish Treaty was officially signed into law on December 6, 1921, in London. This established the sovereign

Republic of Ireland in the south and the British-controlled Northern Ireland in the north.

The treaty left many in Ireland deeply angered and unsettled. A few months after it was signed, this anger would boil over into the Irish Civil War in June 1922, with pro- and anti-treaty forces battling for control of the Irish republic. The war was short-lived: in part because the anti-treaty forces were significantly outnumbered, but also because the pro-treaty forces gained critical support from the British. This allowed the pro-treaty forces to overtake their foes and declare victory in May 1923.

Republican militants, including various strains of the Irish Republican Army active today, can therefore trace their organizational and ideological lineage back to at least the Easter Rebellion of 1916. Ideologically, Irish militants have always opposed the British presence in any part of Ireland and, following the Anglo-Irish Treaty, have also consistently opposed its partitioning. It is for this reason that groups like the IRA are nominally termed *Republicans*: they are ultimately fighting for a single, united Irish republic that incorporates all 32 Irish counties. Organizationally, Republican militants can trace their roots to the Irish Volunteers that formed in 1913 just before the Easter Uprising. During the Irish War of Independence, a portion of the Irish Volunteers broke away and started calling themselves the Irish Republican Army. These individuals were staunchly against the Anglo-Irish Treaty, and they eventually morphed into a guerrilla force that conducted assassinations and other acts of violence aimed at upsetting the status quo and building support for their cause.

Irish Republican militants have therefore been active for more than one hundred years. While levels of violence have waxed and waned, some of the worst violence of the entire conflict took place over a span of thirty years from the late 1960s to 1998. During this period known as "the Troubles," violence was at a fever pitch. By most estimates, more than three thousand individuals were killed and many more permanently injured or scarred.[9] Republican organizations were fighting with the British military and police forces; they were fighting with one another for dominance; and they were fighting against Loyalist, Protestant paramilitary organizations based in Northern Ireland. Loyalist forces have an equally long tradition in Northern Ireland and, as one might expect, their ultimate objective is to resist Republican pressure and to maintain Northern Ireland's status as a British-controlled territory. Hence, they are deemed "Loyalist" for their allegiance to the British government. The Loyalist forces are also overwhelmingly

Protestant due to their historical roots in England and Scotland where many of their ancestors emigrated from during the seventeenth and eighteenth centuries. Since Republican forces are predominantly Catholic, the conflict in Northern Ireland has taken on a religious subtext in addition to its primarily nationalist character.

The Troubles came to an end in 1998 with the Good Friday Agreement, a major political accord between Republicans, Loyalists, and the British and Irish governments. Although the agreement did many things—including devolving additional responsibilities from Great Britain to the regional government of Northern Ireland—it is most notable for establishing a democratic compromise to determine the future of Northern Ireland. According to Article Three, "a united Ireland shall be brought about only by peaceful means with the consent of a majority of the people, democratically expressed, in both jurisdictions in the island."[10] Thus, the majorities of populations in both Northern Ireland and the Republic of Ireland would have to agree to reunification for it to occur. The agreement also recognizes that "while a substantial section of the people in Northern Ireland share the legitimate wish of a majority of the people of the island of Ireland for a united Ireland, the present wish of a majority of the people of Northern Ireland, freely exercised and legitimate, is to maintain the Union."[11] Indeed, the majority in Northern Ireland Support the status quo even to this day.

Although the Good Friday Agreement was hailed as a political breakthrough that precipitated the disarmament and decommissioning of major paramilitary organizations, there are still those who remain unsatisfied. Since 1998, both new and existing armed groups have continued to use violence to undermine peace and achieve their unfulfilled aim of Irish reunification. Since the overwhelming majority of people support the Good Friday Agreement—it earned 71 percent of votes in Northern Ireland and 94.4 percent in Ireland during referendums in 1998—these groups have found little support and have been unable to mount the type of sustained, destructive campaign their predecessors did.

In recent years, dissident violence has reemerged in Northern Ireland. Two factors explain this trend: first, economic stagnation that began with the 2008 recession and has not been fully resolved. And, second, the Brexit referendum of 2016 where citizens of the United Kingdom voted to leave the European Union. A "hard border" that separates Northern Ireland from the Republic of Ireland to the south would anger many on both sides of the border. Not only would trade and business be disrupted, but visits to friends

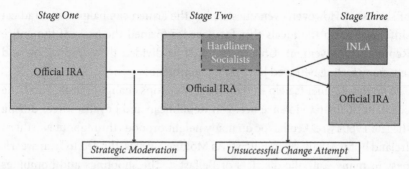

Fig. 3.1 The formation of the INLA

and family would require traversing checkpoints and presenting documents, as happened in decades past. This change would turn a short trip into a multi-hour affair, and it could contribute to a resumption of dissident violence. The specter of violence in Northern Ireland is thus not far removed.

The Irish National Liberation Army

The INLA owes its creation to Seamus Costello. Born in County Wicklow in the Republic of Ireland in 1939, Costello joined both the Irish Republican Army and Sinn Féin at the age of 16.[12] Costello was a bright, charismatic leader and he quickly rose through the ranks of the IRA, becoming adjutant general, chief of staff, director of operations of the Official IRA (one of two IRA successors, the other being the Provisional IRA (PIRA)), and even vice president of Sinn Féin. Costello managed to reach the pinnacles of both organizations and he became a driving force behind a number of key political and military decisions.[13] While he was an ardent supporter of the Republican movement, Costello soon found himself at odds with other leaders of the Official IRA.

Polarizing Events: A Ceasefire and Strategic Reorientation

Costello's disagreements with the Official IRA resulted from various sociopolitical forces and conflict realities that came to a head in the late 1960s. These realities led many in the Republican movement to question whether their armed campaign was hurting rather than helping their chances of success.[14] However, even questioning the armed campaign was a radical shift away from the ideals that for decades formed the basis of the entire Republican movement. Unsurprisingly, this divided the organization and left many feeling betrayed by those in command.

Two developments help explain the leadership's change of heart. First, the constant tit-for-tat violence between Republican and Loyalist forces during the late 1960s wreaked havoc on many neighborhoods throughout Northern Ireland but particularly in Belfast. As Moloney writes, this led to "war weariness in many Catholic districts of Belfast.[15] The shootings and bombings had transformed many nationalist areas into terrifying war zones, where people ran a daily risk of running into gun battles or being caught up in the nerve-jangling bomb explosions." Neighborhoods, families, and individuals were permanently scarred by violence committed during the Troubles, and people were left wondering when it would all end. This view was widely held among the general population but it also trickled up through the ranks of the Official IRA, prompting many to rethink the organization's current strategy.

The second development that contributed to a revision of Official IRA tactics was the imposition of direct rule from Great Britain in 1972. This effectively curtailed Northern Ireland's already limited home rule in favor of direct management from London. It was a momentous event that might seem like a setback to the cause of Irish reunification. In reality, the move was an unequivocal Republican victory—maybe not toward reunification, but definitely toward greater inclusion of Republican voices in Northern Ireland's day-to-day politics. Before the decision, Northern Ireland's regional parliament at Stormont had long been dominated by Loyalists. Protestants outnumbered Catholics in Northern Ireland nearly two to one during the late 1960s and could easily leverage their population advantage into political dominance, marginalizing Catholic and Republican voices. As Bell writes:

Fearful of the larger Irish state to the south, fearful of the minority Catholic population of Ulster, the [Protestant] majority ruled with outward arrogance, determined to maintain their privileges and their way of life. As one Loyalist spokesman indiscreetly admitted, Northern Ireland was a Protestant state for a Protestant people. The Protestant establishment, the Unionist Party within Stormont and the Orange Order without, suspected their minority population to be disloyal, agents of Rome, advocates of the IRA.[16]

So when the British government passed a resolution to transfer control of Northern Ireland to London and away from Stormont, the Republicans were elated: by effectively curtailing the Loyalists' political grasp, many viewed this as a concession by the British government.[17]

While some Republican supporters believed their violent campaign had directly contributed to Great Britain's decision to impose home rule,[18] this policy change ultimately decreased support for additional violence. With war weariness setting in after several years of nonstop attacks, much of the populace and even many Republican fighters believed that their future goals could now best be met through politics rather than violence.

Confronted with the public's war-weariness and the suspension of Unionist rule at Stormont, some Republicans were convinced that a change of tactics was necessary. This idea was further ingrained within the Official IRA after it tried and failed to retaliate against the British army following the Bloody Sunday massacre in 1972, where members of the armed forces shot 26 unarmed civilians and killed 14. The Officials, spearheaded by Costello, decided to retaliate against those responsible for Bloody Sunday by bombing the Parachute Regiment's headquarters in Aldershot, England. The bomb went off as planned but missed its target, killing six cleaning staff and an Army chaplain. This failure and the resulting public backlash "confirmed the fears of those on the [Official IRA's] Army Council who viewed the 'armed campaign' as a political liability."[19] The events at Aldershot bolstered those within the OIRA—mainly the leadership cadre—who favored a political approach over violence, giving them the upper hand in the internal debate.

An Unsuccessful Change Attempt and the Formation of the INLA

Members of the rank and file who opposed the Army Council's moderation soon coalesced around Seamus Costello and Joe McAnn (McAnn was killed by British troops in 1972 just before the INLA broke away). At first, Costello, McAnn, and their supporters "... came together to try and change Official IRA policy" from the inside.[20] Their initial instincts were not to break away as is the case with many splinter groups that fear the uncertain consequences of an organizational rupture. They aimed instead to reason with the Army Council and steer the organization back to its violent roots through existing organizational channels (e.g., votes at the organization's annual meetings). A former associate of Costello's named Gerry Roche noted that "[Costello] wanted to win out through that structure. The last thing I think he wanted to do was to leave the [O]fficials because to splinter weakened your position."[21] They saw the split not merely as detrimental to the Official IRA, but detrimental to the movement as a whole since it highlighted Republican divisions.

At successive Ard Fheis (general organizational) meetings in 1972 and 1973, Costello—with widespread support—put forth a proposal he authored with Sean Garland called "A Brief Examination of the Republican Position: An Attempt to Formulate the Correct Demands and Methods of Struggle." The document outlined the logic and methods of a renewed armed campaign. To Costello's credit, the Ard Fheis passed his plan, but the Army Council vetoed it. Upon hearing the news and witnessing how he could only steer the organization insofar as the Army Council would allow, "Costello realised that this had in fact been the ideal time to leave the Officials. . . ."[22]

Although Costello later said that he should have departed the Official IRA in 1972, he waited another two years while he labored to make his voice heard. However, not everyone in the OIRA was keen on what he was doing and he soon attracted the full attention of the Army Council.[23] They finally voted to expel him outright in the spring of 1974.

Costello's "expulsion only formalised what was already fact—the parting of the way between the revolutionary element and the mainstream reformists."[24] Indeed, the seeds of the INLA had been planted almost two years before as Costello and some of his closest associates began preparing for the split and conducting unsanctioned missions to fund their new organization.[25]

At a meeting outside Dublin at the Lucan Spa Hotel on December 8, 1974, two organizations were officially conceived: the Irish Republican Socialist Party (IRSP) and the Irish National Liberation Army. The meeting went surprisingly smoothly. As Holland writes, "The lack of drama was explained by the fact that the new movement had come together as the result of a gradual process. . . ."[26] These events refute the widespread notion that the breakdown of armed groups is chaotic and haphazard. While the INLA and IRSP worked in tandem to pursue the same overall objectives, the IRSP was tasked with political activities and the INLA with violence and agitation. Some observers in Northern Ireland placed the IRSP and the INLA under the same umbrella and called it the Irish Republican Socialist Movement. But Costello wanted to keep the two organizations separate, at least initially. He hoped to give the IRSP time to mature into a respectable political organization while behind the scenes he would begin acquiring the weapons and money necessary to launch militant operations. Although the IRSP was announced soon after that initial meeting in December 1974, Costello kept the INLA a secret for almost two years, denying its existence at every opportunity.

The INLA's Membership Composition

The subgroup around Costello that eventually became the INLA formed multidimensionally, occupying a large organizational niche at the intersection of ideological and strategic grievances.[h] Although this faction originally coalesced over its opposition to the ceasefire, it soon lost its cohesive advantage by broadening its agenda around socialist objectives, worker rights, and political pressure. As one report from the British Independent Monitoring Commission notes, "The INLA is a very volatile mix of people from many and varied terrorist backgrounds. It has a reputation for extreme violence and internal feuding centered round leadership disputes which regularly lead to fragmentation of the group."[27]

On the political side, the IRSP attracted a "curious mixture of socialists, Republicans and trade unionists, most of whom joined the movement as a protest against the positions adopted by the Provisionals and the Officials."[28] They were attracted to this new organization since they were weary of the OIRA's waning commitment to socialist institutions, believing the OIRA was prioritizing nationalist over socialist objectives. Costello's faction appealed to these individuals by arguing that the "national question and the social question were not to be approached in schematic stages but had to be fought for at the same time."[29] As a result, his faction attracted a wide range of hardcore leftists, all of whom could see their interests being represented in this new group. In particular, they saw Costello's group "as having the potential to become a mass revolutionary party...."[30] Unsurprisingly, this was a far cry from what others in the group envisioned, but it was logical given how the faction early on championed a socialist disagreement with the OIRA.[31]

---

[h] While Morrison (2013) suggests that Costello might have widened the niche intentionally to gain members, he nonetheless capitalized on existing debates within the organization to do so. This underscores how entrepreneurial rebel leaders can indeed manipulate politics for personal gain, but they are still beholden to organizational dynamics.

Problematically for the INLA, there were also major disagreements among the leftists. While the IRSP sought to create a socialist Irish republic that included all 32 counties, it never settled on the exact brand of socialism it supported. Costello and other leaders initially cast a broad net, employing general leftist-socialist rhetoric in their publications. For instance, describing his new group to an Italian journalist, Seamus Costello said that "We are a revolutionary socialist party and our objective is to create a revolutionary socialist state in Ireland."[32] No more specifics were provided. The IRSP also had the tendency of looking past its own identity to focus on the problems with others: "Despite many references to Connolly [a socialist Republican], and to a lesser extent Marx, Engels and Lenin, the politics of the party became defined in terms of differences between its ideology and that of the Official and Republican movements. In other words the IRSP was content to define its political outlook in terms of what it disagreed with."[33] By taking this approach, the group attracted members with wide ranging opinions of their desired ideological foundation. These divisions are reflected in the different communities that were initially drawn to the IRSP:

> The far left were enthused by the emergence of the IRSP seeing it as having the potential to become a mass revolutionary party... People's Democracy [a leftist political organization] also welcomed the formation of the IRSP, and some of its members joined it. Those alienated by the Officials' increasing embrace of Eastern Europe saw the IRSP as potentially "anti-Stalinist." Others hoped it would provide an open forum.... But few within the IRSP, beyond those with a background in the leftist groups, had any knowledge of Marxist ideology.[34]

The negative effect of these divided preferences—even over a single issue— can be seen at the group's first Ard Fheis in December 1975, where eleven council members resigned in protest over the inability to reach a common, coherent doctrine.

Simultaneously, the INLA/IRSP appealed to ardently violent Irish nationalists. Costello's break with the OIRA was not only motivated by socialist, ideological disagreements, but it was simultaneously couched in terms of his antipathy toward the 1972 ceasefire. This attracted some of the most hardline fighters from all of Northern Ireland into his faction. Many of these hardliners were simply criminals who wanted to cause as much destruction as possible, while others viewed the violent campaign as an integral part of the Republican strategy that had to be upheld at all costs. And with both major Republican groups observing ceasefires at the moment (the OIRA and PIRA), this left the INLA as their only outlet. Describing these men, one founding member simply noted that "We were a body of individuals prepared to wage war against the British machine in Ireland."[35]

The hardliners that joined the INLA were only interested in violence and retaliating for the perceived injustices against the Catholic community in Northern Ireland. They had no appetite for politics and little interest in socialist ideology, putting them at odds with other members of the group. As one OIRA member who resisted joining Costello noted, "Many of the people that went with the Erps [IRSP] ... were just keen to get at the Brits and the Prods [Protestants]. They were at heart sectarian. They couldn't resist the temptation to hit out at the loyalists."[36] Another source of hardline recruits came from a withering organization known as Fianna Éirann, an IRA youth wing with a hardline streak. According to a former Fianna Éirann member:

> There was a really militant crowd in the Fianna. I remember one meeting the Fianna was called to in Cyprus Street in 1973. The OIRA quartermaster for Belfast was there. He asked members to tell him how many weapons they had. He couldn't believe what he was hearing. Fianna units reported having heavy machine-guns, explosives, rifles, and handguns. He nearly fell off his chair as he took stock of our weapons. Many of those that later went with the Erps [IRSP] came from the ranks of the Fianna. I was like them at the beginning. All we wanted to do was bang away at the Brits.[37]

Taken together, the INLA's membership and the fault lines that immediately emerged were not a product of chance or of sociopolitical circumstances within Northern Ireland. Other groups did not experience the same lack of cohesion among their members. What plagued the INLA from the start was their broad organizational niche that developed while it was still a faction within the OIRA. This ultimately brought into the group an eclectic mix of members with oftentimes competing, if not fully opposed, preferences for their organizational future. These differences were most stark between the politically minded socialists who saw the INLA as a revolutionary organization, and militant hardliners who viewed the group as the natural heir to violent dissidence in Northern Ireland.

Explaining the INLA's Trajectory

So far I have shown how the composition of the INLA's membership is directly linked to its disagreements with the OIRA and to the broad platform embraced by Costello. But how did the membership eventually influence the group's tactical behavior and its ability to survive?

First, the divergent preferences of INLA/IRSP members generated feuds that tore the group apart and made survival virtually impossible. Fundamentally, members disagreed over the very nature of the armed campaign. Socialist-political members believed it was unnecessary while hardliners believed it was essential. As Holland explains, this manifested itself "between McAliskey on the left wing of the IRSP and the core around Costello in a dispute over the very role of the military wing and the armed campaign."[38] This fault line persisted throughout the entire lifespan of the INLA, and it consistently undermined the group's ability to cooperate and to enforce a coherent strategy.

While Costello was managing fundamental disagreements over the armed campaign, he was simultaneously trying to reign in the most hardline operatives concentrated in Belfast, 100 miles to the north. Those in Dublin were finding it "impossible to exercise any effective control over the Belfast units. They wanted to fight their own war their own way, and to a large extent they did so."[39] Among the group's internal disputes, this was the most dangerous, the most damaging, and the most enduring. The hardliners were eager to wage war against British and Loyalist forces and they planned to do so with our without Costello's support. For example, when Seán Garland, a member of the OIRA's Army Council, was shot and nearly killed in 1974, Costello reportedly asked an IRSP comrade who he thought was responsible. His friend answered "I think we did," and Costello replied in disbelief, "We fuckin' did?" Costello had explicitly commanded the organization to refrain from using violence and his orders had been blatantly ignored. As a result, Costello would soon find himself in a difficult position, negotiating between the socialist-political members, the INLA's violent and uncontrollable hardliners, and his own strategic vision that was situated somewhere in the middle.

As a result of these internal disputes, leaders of the INLA effectively lost control of their organization in only a few years. And with every successive act of unsanctioned violence, group members moved further apart and cooperated even less. By the early 1980s, the virtually uncontrolled violence combined with the Dublin-based leadership's unsettled view toward the violent struggle turned the IRSP/INLA into an umbrella organization of independent units. The IRSP's political ambitions were doomed by the reckless violence of the INLA and their own inability to devise a coherent platform. The situation came to a head when "in early 1985 many spectators believed that both the IRSP and the INLA had ceased to function—both were leaderless and factionalised, the IRSP Cumann [political body] had been dissolved and the INLA Army Council stopped meeting in January."[40] Members of the IRSP were in hiding and the INLA was factionalized in the North, both on the streets of Belfast and even within the jails. This period known as the "INLA feud" witnessed some of the deadliest intragroup fighting and ultimately ended with two separate groups claiming legitimacy over the INLA name.[41]

Second, as anticipated by my theory, the divergent preferences of INLA/IRSP members left the group unable to decentralize. It was forced to rely on strict hierarchy to maintain some semblance of unity. Initially, this was not a problem; when Costello was alive, he consolidated his leadership position to enforce control. Members of the INLA even called him "authoritarian"[42] in reference to his grip over the organization.[i] He knew that strict hierarchy was needed since he could not trust others to act independently. According to his own calculations, a centralized organizational structure was necessary to control the group's competing factions and to prevent outright conflict. It is then no surprise that after Costello was assassinated in 1977, members of the INLA immediately began fighting to take charge, consolidate power, and enforce their own strategic visions.[43]

Third, the INLA was beset by informers that compromised its operations and destabilized its structure. Many of them were enticed by government offers of leniency or cash, while others willingly gave up information to take down opponents within the group. They were willing to cooperate with the government because "there was widespread disillusionment among [members of the INLA] who had grown weary of the divisions and disagreements permeating the organisation."[44] Members who were either

---

[i] Of course, Costello was not entirely successful in this regard. While he did manage to impose a hierarchical control structure on the organization, this did not actually guarantee control. In fact, he still struggled to do so and never fully succeeded.

disillusioned or losing interest were then easy targets for security forces. As one confidential government report from 1979 notes, "The greater risk of arrest and possible conviction will increase the pressure on less committed INLA members and will increase the constraints on activists."[45] Even when the government did not seek out these individuals, they often provided information willingly. As a 1982 article from *The Guardian* simply titled "Feuds breed informers" notes:

> Internal feuding within the Irish National Liberation Army and the result-
> ing defections have led to a series of arrests in Northern Ireland.... It is
> believed that the breakthrough results largely from the decision of a senior
> figure within the organization, who recently survived an assassination
> attempt from a rival internal faction, to cooperate with the security forces.[46]

It is unclear who the informer was, but this example underscores how internal feuding served as a catalyst for defection and infiltration as my theory anticipates. And for the INLA, this feuding was a direct product of the multidimensional disagreements that motivated them to form.

Fourth, while there tends to be an assumption that splinter groups will universally tend toward radicalization, the INLA's trajectory reveals the opposite: their violent behavior flowed logically from the nature of their disagreements with OIRA and from the members they attracted. Hardline recruits flocked to this new organization not by chance, but because it promised to renew the armed campaign. Unsurprisingly, they exerted a steady, radicalizing influence. Sometimes, these members even took it upon themselves to launch their own unsanctioned operations. In one case, as briefly described earlier, Belfast hardliners drove to Dublin and assassinated a member of the OIRA without any approval whatsoever.[47] At first, Costello tried to deny these events. As he vehemently argued (and lied) in January 1975, just months after the IRSP and the INLA formed, "We are not involved in any kind of military action but are solely a political group."[48] But as the number of unsanctioned operations swelled, denial was no longer feasible and he was forced to take ownership of several attacks. Though this may have caught him off-guard, he should have known what to expect: indeed, "Why would gunmen who had grown restless because of the three-year ceasefire join another organization that did not offer them some military role?"[49] Of course, the INLA was not solely compromised of hardliners and its multidimensional agenda simultaneously appealed to more mainstream and less radical individuals as well. Costello tried to strike a balance between their very different interests, but his attempt at moderation partly explains why unsanctioned attacks were so common.

As a result of these challenges, the INLA effectively ceased to exist as a single organization only five years after it broke away from one of the most successful organizations presently operating. It was torn apart by feuding members with fundamentally divergent goals, while the hardline operatives attracted to its anti-ceasefire platform launched increasingly deadly and often unapproved attacks. While this demonstrates the poor foundation for splinter groups emerging from multidimensional schisms, it also shows how their disagreements shape to whom they appeal. Ultimately, leaders of the INLA should have heeded their own advice: as they cautioned in one newsletter, "It is necessary to have clarity about the objectives for which we strive, otherwise the fruits of our struggle could slip to counter revolutionaries."[50]

The Real Irish Republican Army

The Real Irish Republican Army (RIRA) is a splinter organization that emerged from the Provisional Irish Republican Army (PIRA) in 1998. Throughout the 1980s and the 1990s, the PIRA was the dominant and most widely supported Republican organization. The PIRA was created in 1969 when it split from what was at the time the Irish Republican Army.[51] The group was active not only militarily, but also politically through a sibling political organization known as Sinn Féin.

The establishment of the RIRA can, like that of the INLA, be traced to a ceasefire. Since the INLA and RIRA share this important characteristic, we can rule it out as an explanation for their different trajectories. Instead, the RIRA's singular focus on resuming violence is responsible for the group's internal alignment and its long-term survival. While the INLA had a broad socialist and militant agenda, the RIRA's narrow focus meant it attracted a homogeneous group of members who strongly shared the same organizational vision. The resulting preference alignment bolstered its internal cohesion, control, and durability. Likewise, the RIRA's opposition to the ceasefire shaped its strategic course. It attracted hardliners who supported and ultimately launched some of the deadliest terrorist attacks in the country's history.

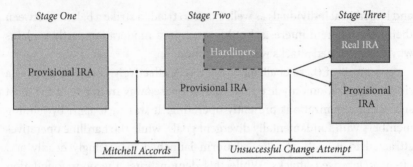

Fig. 3.2 The formation of the RIRA

Polarizing Events: A Unilateral Ceasefire

Despite the PIRA's violent history and decades of dedication to armed struggle, certain events prompted group leaders to rethink their strategic direction in the late 1990s, in a manner very similar to the INLA. During this period, PIRA members began questioning whether continued violence would help achieve their goals. Richard English, an expert on the IRA and the Northern Ireland conflict, identifies three reasons that contributed to this drastic change: the PIRA felt it had reached a military stalemate with the British; PIRA leaders could see definite benefits from ending violence and halting the group's "pariah state status"; and finally, Republicans had come to recognize some of the harsh realities of Northern Ireland that they had previously overlooked or consciously ignored.

First, the PIRA had largely reached a military impasse by the mid-1990s. This is summed up nicely by a Sinn Féin candidate, Martin Ferris, who noted in 1997 that "... Northern Ireland was trapped in a vicious circle. On one hand, the IRA could continue with the war and get nowhere. On the other hand, the British were relentless in their pursuit of IRA volunteers."[52] As this implies, the IRA was failing to achieve meaningful success. There are few victories attributable to the IRA around this time and the North's political situation had barely changed in the past decade. Rather than "[creating] such psychological damage to the Brits that they'll withdraw," the PIRA's violent strategy had devolved into a tit-for-tat cycle where "we can't defeat them in a military sense, no more than they can beat us. So there's kind of a stalemate."[53] This was making it difficult for leaders of the PIRA's to justify the ongoing costs of the armed campaign, and for members of the public to support it.

While members of the PIRA maybe felt like a stalemate had been reached, there is also some indication that the British were in fact gaining the upper hand. This is partly because the British spent part of the 1980s and early 1990s developing a formidable security and intelligence apparatus in Northern Ireland. Combined with the political stalemate and low morale this likely inspired, the British were able to infiltrate the ranks of the PIRA and to monitor and even prevent some of its operation. "By the mid-1980s the capacity of the security forces to constrain Provisional activity through surveillance, arrests and so on was more impressive than it had been in the 1970s."[54] Taken together, many within the Republican movement were coming to question the basic logic of their present strategy.

Second, not only did violence seem unlikely to succeed, but *renouncing* violence seemed promising. Summarizing this opinion at a conference in 1990, a politician from Northern Ireland named John Hume noted that if IRA leaders were willing to abandon armed struggle and have the moral courage to adopt peaceful methods, then "no single act in this century would do more to transform the atmosphere on the island and to begin the process of breaking down the barriers between our people, which are the real problem to this island today and which are the real legacy of our past and which are in fact intensified by the IRA campaign."[55] Ending violence would, according to many observers, usher in a new era where previously unimaginable compromise would finally become possible.[56]

Third, Republicans in the north had come to recognize a few key facts about their situation. Perhaps most prominently, even if the British were driven out of Northern Ireland, the Unionists would still remain. This was not so much of a new revelation as it was a gradual understanding. For decades, the Republicans were so concerned with defeating the British and forcing them out of Northern Ireland that they failed to consider what would happen next. As English notes, "When you're engaged in a struggle, you fight with the basics in mind. It's a united Ireland or nothing; the unionists are basically tools of British imperialism; they don't know what they're doing; they'll come into a united Ireland like sheep once you break the will of the British."[57] But this was not necessarily the case. In some unlikely scenario where the British did leave, the country would still be home to a sizable population of Loyalists who would continue to oppose Irish reunification. To this problem, there was no obvious solution.

In addition, Republican forces liked to talk about the British extracting wealth and resources from Northern Ireland, and they used this fact to generate support from the population. The reality, however, was much different. The British were actually putting more into the North than they were taking out,[58] and these contributions were coming to light during the late 1980s and early 1990s. While Republicans saw the British as foreign occupiers who derived taxes and other economic benefits from controlling the region, Northern Ireland was actually unable to meet its financial needs and Britain had been stepping in to cover the difference. Rather than extracting wealth, Britain was paying to support the regional economy. One observer said that without British support, "...it is wrong to believe that the economy would cease to exist... it would probably step back to closer to third-world levels though probably a 'better-off' third-world type of level."[59] This realization left Republicans increasingly uncertain about the economic success of an independent Northern Ireland, and how they would fund the potential reunification—an expensive endeavor.

Together, these factors made PIRA leaders receptive to the idea of a permanent ceasefire and some sort of compromise with the British. Talks between Northern Ireland, the Republic of Ireland, and the United Kingdom thus began and eventually culminated in the Good Friday Agreement in 1998. The agreement recognized that a majority of people in Northern Ireland desire to remain part of the United Kingdom and it would thus remain so until a majority favored reunification. In other words, it solidified the democratic determination of Northern Ireland's territorial status. The agreement also contained provisions related to the decommissioning of paramilitary organizations and the release of their prisoners. Before these talks could even begin, however, every party involved had to first agree to the Mitchell Principles, a framework established by the US envoy (and former US Senator) George Mitchell. It was these principles, and not the Good Friday Agreements, that are most directly responsible for the RIRA's break with the PIRA. They required each party to agree to the following:

First, To democratic and exclusively peaceful means of resolving political issues;

Second, To the total disarmament of all paramilitary organizations;

Third, To agree that such disarmament must be verifiable to the satisfaction of an independent commission;

Fourth, To renounce for themselves, and to oppose any effort by others, to use force, or threaten to use force, to influence the course or the outcome of all-party negotiations;

Fifth, To agree to abide by the terms of any agreement reached in all-party negotiations and to resort to democratic and exclusively peaceful methods in trying to alter any aspect of that outcome with which they may disagree; and,

Sixth, To urge that "punishment" killings and beatings stop and to take effective steps to prevent such actions.[60]

The first point is the one that most caused division Within the PIRA. Agreeing to exclusively peaceful means "had the effect of demoralising rank and file IRA members whose dedication to armed insurgency against the British in Northern Ireland was proverbial."[61] Many fighters dedicated their lives to the organization and risked everything they had, so they could not fathom how group leaders were willing to disarm for the prospect of talks that could only end with a continued British presence. In addition to their political objections to the accords, other IRA members felt that the act was "an infringement of the organisation's constitution and the negation of the IRA's claim to be fighting a legitimate 'war' against British 'colonial' occupation,"[62] largely because the organization's constitution explicitly forbade such activities short of a complete British withdrawal. Consequently, many PIRA members opposed the Mitchell Accords and felt their leaders were betraying their organization's fundamental principles and traditions.

Formation and Exit of the RIRA

Debates over strategy during the 1990s led to the establishment of two oppositional camps within the OIRA. The majority, headed by Gerry Adams and Martin McGuinness, favored compromise and de-escalation. Hardliners were represented by Michael McKevitt, Seamus McGrane, and "Frank McGuinness," a pseudonym for the PIRA's top bombmaker. By the time another ceasefire was approved in 1998 in accordance with the Mitchell Principles, hardliners like McKevitt were already preparing to depart. As far back as 1994, McKevitt had been:

[assembling] a group of confidantes who would meet in secret to discuss IRA policy and the future direction of the underground army. This select group talked a good deal among themselves; the common denominator was that none trusted the Army Council. All watched [Gerry] Adams' pronunciations with frightening attention, analysing his comments and public statements with microscopic interest.[63]

As with the INLA, this split was no haphazard event.

Just as Costello was central to founding the INLA, McKevitt was to the RIRA. An IRA member for most of his life, he joined the group as a teenager and rose through ranks. He came to be known as a lethal, notorious operative. By the end of his tenure in the PIRA, he had become quartermaster general and was tasked with arms procurement, training, and weapons storage. These skills would prove to be quite useful, helping his new group be especially productive and successful in its first year of operation.

McKevitt's ambitions for his breakaway group were straightforward: he "aimed to uphold any uncompromising and uncompromised Irish Republicanism, and to oppose anything emerging from the 1997 party talks that should fall sort of Irish unity and independence."[64] The core issue for McKevitt and his supporters was the ceasefire and the decommissioning of IRA weapons, which to them was antithetical to the PIRA's constitution. As far as they could tell, the upcoming talks (and any permanent ceasefire) would never lead to British withdrawal from Northern Ireland. McKevitt and his factional comrades therefore strongly opposed the entire peace process. Instead, they remained committed to their militant Republican traditions and refused to believe that their struggle had been in vain.

In another similarity with Costello, McKevitt initially hoped to steer PIRA policy from the inside, making the case against the proposed ceasefire. However, it soon became obvious that the PIRA's leaders had outmaneuvered the hardliners and would move forward with the Mitchell Accords. McKevitt's subgroup decided to break with the PIRA in November 1997 and form their own organization—the "Real IRA"—where they could oppose the ceasefire.

The RIRA's Membership Composition

The RIRA was born out of a clear, unidimensional disagreement with the Provisional IRA: its members opposed the 1998 ceasefire with British and Loyalist forces and, more generally, the abandonment of the revolutionary armed struggle in pursuit of a reunited 32-county Ireland. As they repeatedly argued, "Our goal is the same as the IRA's has always been—to force a British withdrawal. We're no different than the men and women of 1916, 1919, or 1969."[65] While PIRA leaders were willing to negotiate with the British and disarm for the sake of a negotiated compromise, those in the RIRA Army Council "cannot envisage a ceasefire in any circumstances other than in which a declaration of intent to withdraw from the occupied Six-Counties is made by the British Government."[66] Since the PIRA's original constitution strictly forbade decommissioning short of a British withdrawal, McKevitt and his followers felt "they had remained faithful to the IRA's Constitution; [and] they were the Real IRA."[67] Hence, the meaning of their name.

The RIRA's goal was thus straightforward and singular: to disrupt the peace process by conducting attacks throughout Northern Ireland.[68] The RIRA "had a vision or belief in pursuing a 'military' campaign until they achieved a united Ireland," while in contrast "the Provisionals appeared to be edging toward an interim compromise of some kind, prompting traditional Republican fears of a sell out."[69] The British government's assessment was equally clear: the RIRA "was formed by defecting members of PIRA who were opposed to the 1997 ceasefire and later to the Belfast Agreement."[70] The RIRA planned to launch violent operations to derail the existing negotiations and then pursue the ultimate Republican objective: the reunification of Ireland. While the PIRA was willing to observe a ceasefire for the sake of negotiations, the only event that could possibly precipitate a RIRA ceasefire would be "a declaration of intent by the British to withdraw their military presence from Ireland and to cease all parliamentary activity here."[71] McKevitt and other RIRA founding members basically had no interest in politics.[72]

Nonetheless, McKevitt and others still recognized that they could benefit from a legitimate political organization that could debate Sinn Féin in public while they concentrated on violence. For this reason, the RIRA split "coincided with the separation from Sinn Féin of the 32 County Sovereignty Committee (now 32 County Sovereignty Movement), a group commonly thought to be the political voice of RIRA."[73] The 32 County Sovereignty Committee started out as a bloc within Sinn Féin, but within a year the members had been expelled from the organization after they were physically barred from entering a Sinn Féin annual meeting in 1998. From this point on, they formed an independent organization, changing their name from the 32 County Sovereignty *Committee* to the 32 County Sovereignty *Movement*.[74] Indeed, this is quite similar to how the INLA formed with their political wing, the IRSP.

The new party (abbreviated as 32CSM) was comprised of disaffected Sinn Féin members who, like their RIRA counterparts, were unenthusiastic about the prospects of a political settlement. The 32CSM was initially led by a formidable Republican with equally formidable credentials: Bernadette Sands-McKevitt, whose brother was Bobby Sands, famous for leading and then perishing in the INLA's 1981 hunger strike. The group's first chairperson was Francie Mackie, well known for his hardline views. As Frampton notes, "Mackey's uncompromising message of support for 'armed struggle' and his undiluted vision of Irish 'sovereignty' came to define the 32CSC/32CSM...."[75]

Members of the 32CSM were therefore very different from those who remained in Sinn Féin and even from members of most other political parties (including the IRSP). In fact, they were barely interested in politics at all. "This was not a political party in the traditional sense, but rather its members viewed themselves, in the formulation of Sands-McKevitt as 'watchdogs over Ireland's sovereignty.'"[76] The party's core ethos was summarized by Sands-McKevitt in a radio interview from December 1997. She ended the conversation by reading a passage from her brother's diary during the hunger strike:

> I'm standing on the threshold of another trembling world.... I am a political prisoner.... I believe in the God-given right of the Irish nation to sovereign independence and the right of any Irish man or woman to assert this right in armed rebellion ... there can never be peace in Ireland until the foreign oppressive British presence is removed....

As this indicates, even though the 32CSM was considered the political arm of the Real IRA, its solutions were far from political. As Michael McKevitt much later confided to an FBI agent, "[32CSM] were all military people and were put there for that purpose to keep army politics in the hands of the military."[77]

The RIRA and the 32CSM therefore had very similar if not complementary goals. And their clear, concise message attracted a group of dissidents with highly similar preferences and organizational objectives. As Mooney and O'Toole argue, "McKevitt had conducted a relatively successful recruitment drive. He amassed a formidable force of volunteers. The recruits were hardline republicans; they saw the IRA not as a political organisation but as a religion."[78] Members of the 32CSM were likewise opposed to negotiations with the British. Their mission, however, was to support RIRA through non-violent means, working in tandem with the armed campaign. Because these individuals held the same basic vision for their organizational trajectory, internal disputes and defection were initially minimized. The harmonious relationship between the RIRA and 32CSM stands in stark contrast the contentious relationship between the INLA and its political wing, the IRSP.

Explaining the RIRA's Trajectory

The disagreement that led to the RIRA's formation helps explain their behavior and durability. As I have discussed, the RIRA's strong, vocal opposition to the PIRA's ceasefire made the organization's goals abundantly clear, so they attracted a core group of ardent hardliners disaffected with the PIRA's softening approach. These individuals lent their support to leaders like Michael McKevitt, who held equally hardline views, and they were attracted by the dissident, violent narrative that motivated the RIRA's departure. They saw the RIRA as the ideal outlet that would embrace their radical vision. Many of these recruits "studied Irish history and would often refer to men like Padraig Pearse, the leader of the 1916 rebellion who sacrificed his own blood for his dream of a United Ireland."[79]

Whereas the RIRA's vocal disagreement with the Mitchell Accords' conciliatory measures guaranteed that hardline militants would be drawn to the new organization, it also meant that only a very specific type of recruit wanted to join. Individuals joining the RIRA were not interested in conventional politics and did not have a socialist agenda. They joined to reignite the militant brand of Republicanism they believed was key to achieving a unified Ireland. The RIRA made it clear that nonviolent approaches were futile, and the resulting preference alignment within the organization bolstered its durability in several discrete ways.

First, the group was able to adopt an organizational structure with devolved authority and a parallel shadow council. The council was ready to "run things in case the main players went inside" and this allowed them "withstand arrests and still maintain violent activity."[80] The establishment of a shadow council was a significant move that supported the organization's ability to survive. Indeed, "If a group is highly institutionalized and has clear lines of succession, then the loss of a leader would presumably be less likely to cause major changes in its direction."[81] It is also less likely that the organization will descend into chaos as potential leaders vie for power in the event of significant arrest or death. This is not to say that organizations lacking clear succession plans are doomed, but rather that "a clear line of succession [facilitates] success."[82]

The RIRA also benefited from a decentralized organizational structure that delegated autonomy to units operating throughout Northern Ireland. Numerous sources suggest that the Real IRA utilized an embryonic, cell-like structure. Under this setup, major decisions were made by a governing body consisting of an Army Council and an Army Executive, but individual operations were planned and carried out by smaller Active Service Units.[83] As one Independent Monitoring Commission Report notes, "The RIRA lacks an organised structure so that individual units have a considerable degree of autonomy. There is little central strategy although there is input from leadership figures in terms of authorising or overseeing attacks."[84] This sort of compartmentalization provides significant benefits to militant groups largely because it helps minimize the effects of government infiltration and operative defection—something that all Republican organizations struggled with over the years.

What enabled the RIRA to decentralize its operational command and create a shadow council was its strong internal preference alignment. McKevitt knew that the militants under his control could be trusted with a certain degree of operational autonomy. He also knew that he could create the backup council that would sustain the group's mission in his absence. Indeed, as Shapiro notes, "The more that the preferences of principals and agents in terrorist groups diverge, the worse it is for the principals to have operatives doing what they want."[85] The RIRA had little to lose from devolving autonomy since the organization's agents could be trusted to carry out operations without strict oversight. Not surprisingly, this type of transformation would have been unthinkable with the INLA, a group that remained hierarchical throughout its history and that even witnessed deadly feuds between competing leaders. Ultimately, McKevitt's ability to structure the RIRA in this way helped it to withstand infiltration and arrests by the British government.

Second, the types of arguments that did arise within the RIRA were generally minor compared to the arguments within the INLA. They tended to be about degrees of strategy and not necessarily alternatives, and these feuds did not result in direct intragroup conflict for dominance. Consider one particular feud that beset the RIRA at its first organizational meeting:

There was deep division about how best to proceed. "McGuinness" wanted to adopt a new approach and was very clear sighted about the situation. As the republicans listened with placid attention, he said a murderous campaign against British soldiers and police would be the best approach. "McGuinness" argued against using car bombs to destroy towns and commercial targets. Such attacks were useless and didn't advance the cause . . . Campbell [another senior RIRA member] thought otherwise. Large bombs made an impact and sent a clear message to the British government; bombings made Northern Ireland ungovernable. He also argued that there was a greater chance of success with a car bomb than trying to shoot a British soldier, or shoot down a helicopter.[86]

The matter was discussed and resolved by the Army Council. "Campbell was permitted to run whatever military campaign he felt was necessary" and " 'McGuinness' was satisfied once there was a relentless onslaught against the British; he was content to let the matter rest."[87] Compared to what the INLA experienced in its first year, including the resignation en masse of much of their governing council, this can barely be considered a division. This example also shows how individuals with radical preferences can indirectly shift group strategies toward the extreme as I theorize in the previous chapter. When McKevitt allowed both Campbell and McGuinness to conduct their own operations and essentially run their own sub-units within the RIRA, he appeased his more radical members to maintain cohesion. However, this decision also guaranteed that his group would be responsible for even more destructive acts of violence.

Third, and finally, McKevitt's strategic disagreement with the PIRA's leaders provides insights into the organization's trajectory. The RIRA's initial break with the PIRA over the ceasefire virtually ensured that their group would undertake violent actions. As Mooney and O'Toole note, "With no political agenda other than to collapse the ongoing peace negotiations, there was no doubting the threat the RIRA represented."[88] Just as with the INLA, the RIRA's vocal opposition to the ceasefire attracted some of the most ruthless Republicans. These individuals were looking for an outlet to conduct armed operations against British and Loyalist targets. The RIRA thus attracted "battle-hardened terrorists in its ranks who are unlikely to be deterred from future violence by the Omagh tragedy"—even though Omagh, a bombing in 1998 that killed 29 and injured over 200, even forced McKevitt and other members of the Army Council to rethink their violent strategy. It was not inevitable that the RIRA would therefore develop into the relentless militant organization it became, but it was a logical outcome given the preferences of individuals who joined the group when breaking from the PIRA.

Alternative Explanations

Could other factors explain differences between the INLA and the RIRA? Of course, factors beyond a group's organizational dynamics will affect how it behaves. No group is an island unto itself and all operate within the context of a broader sociopolitical environment that they shape and are shaped by. Even accounting for some of these factors, a splinter group's fundamental trajectory is inextricably linked to how and why it initially formed.[89]

I consider three other potential explanations for the INLA and RIRA's trajectory: leadership decapitation and the loss of key figures, competition between each group and its parent organization, and differences in their sociopolitical environments. While each of these explanations contributes to a more comprehensive understanding, none provides a better explanation for each groups' development and demise than the explanation based on their initial formation.

First, one could argue that losing Seamus Costello—the INLA's charismatic founder—to assassination only a few years after the group's emergence precipitated its downfall and explains its inability to survive. Indeed, "headhunting," or taking out a group's leader, is a common counterinsurgency and counterterrorist strategy that can be effective.[90] This explanation for the INLA's demise is insufficient: for one, the INLA's troubles were evident from the start and the heterogeneous mix of INLA recruits were feuding *before* Costello was assassinated. For instance, there was a mass resignation of leaders at the first INLA political meeting in 1975. This implies that the forces responsible for the INLA's demise were already present. Interestingly, some have speculated that Costello's death might have actually helped the organization as it relinquished his strict control and temporarily brought members together.[91] As for the INLA's operational profile, it could also be argued that losing Costello explains why the group quickly spiraled into violence. Perhaps Costello, with his strict control of the organization, was able to contain the group's most ruthless elements, and once gone there was little holding them back. As before, this appears compelling but is in fact insufficient because Costello never fully wrested control of the hardliners even while alive.

It is also important to note that leadership turnover was something that affected all Republican organizations and not just the INLA. The RIRA's charismatic leader, Michael McKevitt, was imprisoned only five years after the group formed, so both the INLA and RIRA were forced to deal with leadership turnover relatively early on.[j] Here, however, we can see the benefit of decentralization: the RIRA was prepared and plans were in place for others to take over, ensuring a relatively smooth transfer of power.[92] And, the RIRA's operational profile before McKevitt's arrest was largely similar to that after his arrest. Aside from an operational lull following the public backlash against the Omagh bombing, no fundamental strategic or tactical changes were noted, and patterns of lethality were largely unchanged. Yet, McKevitt's arrest nonetheless strained the organization. Along with inconsistent statements regarding a ceasefire, the RIRA soon experienced a split of its own with Óglaigh na hÉireann (ONH) coming into existence. While not much is known about ONH, it is telling that the RIRA nevertheless persisted, and that the ONH even remained loyal to McKevitt while he was imprisoned.[93]

A second plausible argument regarding the differences between INLA and RIRA has to do with intergroup competition. There was significant feuding between the INLA and its predecessor, the OIRA,[94] that created an especially inhospitable environment. On the other hand, the RIRA may have faced a more lenient operating environment as its parent group (the PIRA) was disarming under the auspices of the British Independent Monitoring Commission. Taken together, these factors could explain why the RIRA survived while the INLA struggled. While fighting between the INLA and OIRA was certainty intense,[95] any argument that this intergroup fighting harmed the INLA is belied by the fact that INLA thrived during Nothern Ireland's period of deadliest violence. As one declassified British intelligence report notes:

> The Officials were determined not to let the IRSP grow without a struggle.... It is indicative of the support for the IRSP that despite constant harassment from Officials the party quickly blossomed. Soon after its foundation the IRSP made deep inroads into Official IRA membership in Northern Ireland, particularly in Belfast and Londonderry, and by the Spring of 1975 it was claiming a membership, almost certainly exaggerated, up of to 800.

[j] McKevitt was arrested following the RIRA's Omagh bombing.

Far from faltering, Costello's organization emerged relatively unscathed.

Like the INLA, the RIRA also experienced retaliatory violence and public condemnations from its parent organization.[96] While the PIRA was technically committed to disarming, few believed it was fully carrying through on its promise. One member of the British army told *The Guardian*, confidentially, that in 2000 "the [Provisional] IRA's capability is higher now than it was two years ago," before decommissioning began.[97] Had the PIRA fully disarmed, the RIRA would have had an operational advantage. Yet, the opposite is true, and the PIRA was committed to undercutting its splinter.[98] As Silke notes, "Allowing existing rival Republican groups to expand or, more seriously, allowing entirely new groups to emerge posed a serious threat to PIRA supremacy in nationalist areas."[99] These dynamics help to explain why, in 1998, "members of the RIRA received a 'knock on the door' from the Provisionals who informed them, in no uncertain terms that if they 'stepped out of line again, [they would be] shot.'"[100] It was not only words that the PIRA directed against the RIRA. In 2000, the PIRA shot and killed the RIRA's local commander in Belfast, Joe O'Connor.[101] Thus, when evalutating why the RIRA and INLA had different organizational trajectories, there is little evidence that the RIRA's relationship with its parent group was any better than the INLA's.

Intergroup competition and direct fighting could also affect each group's level of violence. Since new organizations pose a threat to the dominance, reputation, and credentials of other organizations, each may up the ante of violence to prove themselves.[102] While it is plausible for increases in violence to occur following an intergroup dispute, the experiences of the INLA and RIRA provide little evidence of this behavior. As I have shown, there were discrete factions within almost every Republican group, with some wanting to use more violence and others wanting to use less despite *identical* operating conditions. The theory of outbidding, on the other hand, would posit an almost monotonic effect based on the number of groups in the local environment, and it cannot account for this clear variation among groups' strategic choices.

Third, and finally, one cannot separate armed groups from the environments in which they operate. Maybe conditions were simply more hospitable for the RIRA in 1998 than they were for the INLA in 1972, which might explain why the INLA found it so difficult to persist as a coherent organization. The opposite seems to be true.

To begin with, the INLA was able to attract more members than the RIRA did, indicating greater public support. INLA membership eventually reached nearly 800, while membership for the RIRA never reached more than 150,[103] 200,[104] or perhaps "several hundred" at most.[105] In addition, public opinion was shifting against the armed campaign over the course of the conflict, meaning that the INLA generally faced a more favorable operating environment. The general public in Northern Ireland went from somewhat supportive or neutral in its attitude toward armed conflict during the 1970s to largely against it in the early 2000s.[106] For the RIRA, the public's opinion of the armed campaign was already more negative than when the INLA formed. It especially shifted against armed conflict, and against the RIRA itself, following the Omagh bombing in 1998.[107] Tongue also notes— in an article tellingly titled "No-One Likes Us; We Don't Care"—that around this time, the Republic of Ireland was united in opposition to the armed campaign.[108] International public opinion likewise shifted against Republican militants and specifically the RIRA. Following the September 11, 2001 attacks in the United States, the US government declared the RIRA a terrorist organization, prohibited fundraising for the relatives of Irish dissident prisoners, and shut down the RIRA's websites.[109] These were meaningful changes that meant that "Republicans outside the mainstream could no longer look towards America for substantial funding."[110] Taken together, the loss of local and intentional support meant conditions faced by the RIRA were more inhospitable than those faced by the INLA.[111]

Could differences in the sociopolitical context also influence patterns of radicalization, explaining the RIRA's consistent strategy and the INLA's more erratic approach? It is difficult to establish a clear causal link. On the one hand, pressures to engage in political negotiations and disavow violence should have been felt most strongly by the Real IRA. Against the backdrop of the popular 1998 Good Friday Accords, and fatigue from 30 years of the Troubles, the RIRA would be expected to face significant obstacles in maintaining a violent strategy. Yet RIRA members consistently doubled down on violence and only relented when imprisoned. The INLA, on the other hand, garnered more support for its violent strategy in the 1970s. Support was not overwhelming, but it was easier to justify the armed campaign following the Bloody Sunday Massacre that occurred only two years before the INLA's emergence. And while some research suggests that the psychological consequences of societal alienization can strengthen in-group unity by reinforcing an us-versus-them narrative, both the INLA and RIRA were in the clear minority and perpetually fearful of arrest.[112]

In sum, although the RIRA and the INLA certainly experienced distinct operating conditions, their divergent trajectories cannot be explained by the groups' external environment. Instead, they seem better explained by the groups' internal dynamics that can be traced back to their initial formation. It was their membership compositions and organizational visions that propelled them along distinct trajectories. While the INLA and RIRA both navigated some different, and some similar, challenges, one can clearly trace the enduring impact of some of their earliest actions.

Conclusions

In this chapter, I show how the multidimensional disagreements that motivated the INLA to form generated a heterogenous, divergent membership base that precipitated its eventual decline, and how the RIRA's unidimensional, strategic split produced homogeneity and unity that bolstered the group's chances for survival. While the Real IRA was more steadfastly radical and determined to use violence, I show how neither group's actions were inevitable, but rather, they flowed logically from the types of members that were attracted to both organizations.

The INLA's ability to survive was undermined by internal conflicts and by its hierarchical structure that sought, in vain, to exert control. Maintaining a top-down, centralized structure left the group vulnerable to arrests, infiltration, and power struggles—all of which were already more likely due to the disagreements among its members. The Real IRA, on the other hand, capitalized on its internal consensus to decentralize its operations, create a cell-structure that was difficult to penetrate, and establish a shadow leadership that could take over in an emergency. This ensured the RIRA would continue even after the current leaders are gone. The RIRA was able to decentralize precisely because its operatives could be trusted and because there was minimal disagreement between them. There was already a much lower chance of feuds and defection within the RIRA, because of the alignment of members' preferences, and the group was more resilient to the pernicious effects of feuds and defections due to this flatter organizational structure.

The operational profiles of the INLA and the RIRA also were shaped by the preferences of individuals attracted to each organization. Tactical and strategic hardliners were drawn to both groups in response to their anti-ceasefire, anti-compromise approach. The INLA's tactical behavior was a mix of highly radicalized, deadly operations like the assassination in 1979 of Airey Neave, the British Secretary of State for Northern Ireland, and more straightforward political maneuvering. The INLA's political engagement was led by core members who were drawn to the organization due to their socialist and political commitments, and these members had little appetite for violence. They were more interested in uniting the working class for a potential revolution.

The Real IRA's singular focus on violence and its strict opposition to political compromise attracted a more focused core of extremists with decidedly hardline preferences. Unlike those who joined the INLA, members of the RIRA were uninterested in politics and were united by their shared desire to reunite Ireland by force. Consequently, the RIRA's behavior was consistently radical. Not surprisingly, the Real IRA is responsible for the deadliest attack of the Troubles, the Omagh Bombing, which left 29 dead and over 200 injured.

Ultimately, the evidence presented in this chapter shows how the behavior of breakaway splinter groups is strongly linked to the disagreements that motivated them to form. And more specifically, comparing the Real IRA with the INLA demonstrates that a group's membership composition exerts important effects. Their preferences determine the types of operations that groups will launch. One of the most important tasks for leaders of militant organizations is to preserve unity, so leaders will often seek to appease their fighters, and especially the most hardline elements. With both the INLA and the RIRA, Costello and McKevitt ramped up or merely accepted more radical violence out of concern for organizational unity. Preferences also shape group decision-making in more direct ways by skewing internal discourse. For instance, Real IRA discussions were dominated by hardline perspectives since they lacked more moderate voices.

While members' preferences reveal how groups will act, they also shed light on their ability to survive. When armed groups are comprised of members with very different goals and perspectives, it will be difficult to maintain cohesion and survive. This was the case with members of the INLA who exhibited wildly different strategic visions. Some were staunch militants, while others were opposed to violence and sought victory through more conventional political means, and others still through a socialist uprising. And all thought that joining the INLA made sense since the INLA championed a wide range of disagreements with its parent organization. However, these individuals never saw eye to eye, and their antipathy and disillusionment led to feuds, defection, and other unsanctioned behavior. The RIRA, conversely, benefited from having a core group of similarly minded militants. They experienced greater cohesion and were able to decentralize into a cell-like structure. As these cases show, internal cohesion is critical to the survival of armed groups.

Overall, the INLA and the RIRA demonstrate the explanatory power of my theory. We can trace their behavior and survivability to their reasons for forming. But more specifically, the fine-grained case study reveals the causal mechanisms that underpin this connection. Their reasons for forming strongly influenced their membership composition, which in turned shaped how they behaved and whether they could survive. I continue to test this proposition in the following two chapters. In Chapter 4, I present statistical evidence showing how my theory operates in different countries and over time. Then, in Chapter 5, I study the formation of the Islamic State from the ranks of Al-Qaeda, and I show how my theory even operates in a less typical case of organizational splintering. All of this evidence makes it clear that splinter groups are more complex than commonly understood, and that their trajectories are linked to their initial motivations for breaking away.

4

Statistically Evaluating How Splinter Groups Emerge and Behave

Introduction

In the previous chapter, I provide qualitative evidence to support my theory. I trace the experiences of two Republican splinter groups that formed in Northern Ireland during the Troubles. Since they share so many similarities—e.g., they emerged from well-known and successful parent groups, they broke away during ceasefires, they had charismatic initial leaders, and so on—one might assume they would follow similar trajectories. They had markedly different experiences, however, and I show how these experiences are in fact linked to their different motivations for forming.

While my study of Northern Ireland provides compelling evidence in favor of my theory, it is limited to one country and two organizations. In this chapter I provide quantitative evidence showing that it also operates more widely. To do so, I collect a new data set that details how 300, randomly chosen armed groups formed and splintered. I use this to statistically evaluate how breakaway groups behave and to what extent they survive. Simply put, the results reinforce what I have found so far: across countries and over time, there is a strong link between how splinters form and their ultimate trajectories. More specifically, strategic disputes generate splinter groups that are the deadliest of all, while those emerging from unidimensional ruptures are especially resilient. These analyses also reveal that splinter groups, as a whole, defy any simple categorization: they do not show any distinct patterns of survival or behavior, nor do they emerge at higher or lower rates from specific types of organizations. Differences only arise when one takes into account their motivations for breaking away.

In what follows, I begin by describing how I collected this new data set before exploring basic questions about the frequency and likelihood of organizational splits. From here, I conduct more rigorous statistical tests of my theory, evaluating the radicalization of splinter groups first and their survivability second.

Divided Not Conquered: How Rebels Fracture and Splinters Behave. Evan Perkoski, Oxford University Press.
© Oxford University Press 2022. DOI: 10.1093/oso/9780197627075.003.0004

Building a New Data Set

To make the statistical evaluations as robust and as generalizable as possible, I created a time-series, cross-sectional data set of 300 randomly chosen armed groups. In effect, this details a sample of terrorists and insurgents over their entire existence. By randomly choosing these groups, I can ensures that my sample is not biased in favor of well-known or especially deadly organizations, and that it contains a mix of splinter and nonsplinter groups.[1] This last feature allows me to compare how splinters behave relative to one another, but relative to nonsplinter groups as well.

With these considerations in mind, I began with the universe of groups identified by the Global Terrorism Database (GTD). This includes over 3,500 unique actors. I then removed attacks by individuals, by unknown perpetrators, and by groups that are missing information.[a] Since my theory is specific to armed *organizations*, it should not apply to individual actors or unorganized groups of students or protesters that do not experience the types of group dynamics I am interested in studying.[b] Second, I also removed organizations that never managed to kill a single individual since this project is concerned with the evolution of nonstate actors that intentionally use violence against civilians to further their political, social, or economic agendas. Groups that do and do not kill civilians likely differ in many ways and may experience different processes of organizational breakdown. For instance, they may be especially prone to disagreements over the use of force.

After removing unwanted groups from the GTD, I merged the remaining information with data from the RAND-MIPT project on Terrorist Organizational Profiles. This database details the ideology, goals, home country, maximum size, and other relevant characteristics of armed groups. It is important to include some of these characteristics in my statistical analyses to be confident that variation in survival and radicalization is due to how splinter groups formed, and not one of these other features. Following the data merging process, I was left with 850 organizations that were present in both the GTD and the RAND-MIPT databases. In the final

---

[a] This includes "Long-Time Muslim Immigrants from Bangladesh," "Vietnamese Detainees," "Insurgents," "Individuals claimed to be Policemen," "Communists," "Guerrillas," and so on.

[b] And, while the properties of unclaimed operations and those committed by unknown groups are important, without more information I cannot study these groups.

step, I used a random-number generator[c] to select 300 organizations that I would study.[d]

Once the random sample was generated, I meticulously researched all 300 groups to determine three things. First, how did the group form? To qualify as a splinter, the organization must form when a minority of group members break away to fundamentally alter the status quo and to form their own, independent armed group. Second, if a group was formed by splintering, I then researched when the split occurred, why it occurred, and from which parent group it departed. In the previous chapter, I write that armed groups usually experience disagreements across three, non-exclusive issue areas: ideology, strategy, and leadership. I used the following criteria to code these issues.[e]

Ideological splits are sparked by disagreements over worldviews or religious, political, social, or economic goals (and not how they are being pursued). These can be differences over existing goals and worldviews, such as when certain leaders of the Southeast Asian Islamist organization Jemaah Islamiyyah charged their peers with "having Shi'ite and Sufi tendencies and therefore of having strayed from Salafi teaching."[2] However, ideological disagreements can also stem from changing goals and worldviews. Within Germany's Red Army Faction, "ideological changes failed to galvanize support, instead confusing and turning away potential supporters because of the frequent shifts in objectives."[3] Ideological disagreements often concern implementation as well—for instance, how socialist institutions or Islamic caliphates will function. Questions of ideological implementation frequently divided the Basque terrorist group, ETA, and factions repeatedly formed around competing visions for the eventual Basque state.[4]

Type of Split	Description
Ideology	Disagreements over selection, dedication, implementation, interpretation of goals and worldviews
Strategy	Disagreements over the use of force including ceasefires, negotiations, targeting, tactics
Leadership	Disagreements over power, control, competency, trust, betrayal

[c] The number 300 was chosen to balance the feasability of carefully researching all organizations with necessary statistical power.

[d] Owing to their inclusion in two different data sets, there is a possibility that my sample of armed groups is somewhat biased towards more lethal, more durable, and more well-known organizations. However, this should actually provide a harder test of my theory since these minor selection effects might result in less variation across my two dependent variables (radicalization and survivability), making increased rates of both harder to identify.

[e] Each of these are coded dichotomously, or in effect, yes or no.

Strategic splits are driven by disagreements over the use of force. This might include targeting choices, whether to escalate or pause violent operations, or whether to accept a ceasefire or negotiated settlement (thereby ending violence). These are common among armed groups, particularly when governments make conciliatory offers. For instance, several factions broke off from the Palestine Liberation Organization during the 1970s when it entered into negotiations with Israel.[5] However, strategic splits also occur independent of government intervention. Both Palestinian Islamic Jihad and Hamas broke away from the Muslim Brotherhood over "the value of violent versus nonviolent resistance."[6] Likewise, Tanzim Qaedat Al-Jihad split from Jemaah Islammiya in the early 2000s when hardline JI operatives like Noordin Top wanted to continue bombings, while others wanted to focus on ministry activities in the southern Philippines.[7]

Finally, there are splits driven by leadership disputes. Rather than stemming from strategic or ideological differences, these are centered upon control, power, competency, or betrayal. In the tense, precarious existence of armed groups, it is no surprise that these sorts of disagreements are common. For instance, numerous groups have split when there is an opening for a new leader to emerge, prompting those in the group's upper echelons to fill the void. These opportunities often follow the death of a previous leader. As noted earlier, this was a concern when Al-Qaeda lost its presumptive heir, Hamza bin Laden, a few years after his father's death,[8] and then another central figure, Asim Umar, not long afterwards.[9] But leadership fractures do not only occur when individuals are killed or captured. Fatal Al-Islam, a Palestinian Islamist organization founded in Lebanese refugee camps, split in 2006 when "Al-Abssi, a senior leader, felt betrayed after [other] leaders handed over two of his men to Lebanese intelligence."[10] Splits within Egypt's Muslim Brotherhood were similar. "As the [Muslim Brotherhood's] institutions became dysfunctional, the question of who should lead the group became a central debate, leading to the rise of two rival camps...."[11] Notably, these leadership disputes were largely products of internal forces with outsiders relegated to supporting roles.

To understand why splits occurred among the 300 randomly selected armed groups, I consulted a wide range of information. Sources included written histories of individual organizations, news stories, analyses of particular leaders and individual conflicts (e.g., accounts of the Troubles in Northern Ireland), published academic research, and unpublished dissertations and manuscripts. For news articles, I searched Factiva and LexisNexis for relevant information about each group, looking for words like split, splinter, break, disagreement, fracture, depart, and any iterations of each (e.g. splinters, splintered, and so on). Lastly, and when needed, I conducted more general searches through Google, JSTOR, Google Scholar, and others for references to particular cases of group breakdown. At minimum, I searched for two different accounts of why a particular split took place. I then coded every disagreement noted in the narrative according to the categories outlined above.[f] This yields three binary indicators: one each for strategic, ideological, and leadership disputes. After coding the specific motivations underlying an organizational split, I then coded—again with a binary indicator—if it was unidimensional, where sources confirmed that a single issue motivated the split; or multidimensional, where it was caused by two or more disagreements. I was able to understand the vast majority of organizational fractures using these methods.[g]

Evaluating the Data Set

Before delving into patterns of survival and radicalization, some useful inferences can be gleaned from the aggregate data.

First, how common is organizational splintering? There are two ways of looking at this: we can examine how many of the 300 randomly selected groups formed by breaking from another, preexisting organization. Figure 4.1 (left) shows that 28.33% of organizations in this sample (85 of 300) formed not from the ground up, but by splintering from an existing group. Another way to approach this data is by studying groups' "family trees." In other words, every time a splinter emerges from one of the original 300, we add it to the total. As we can see in Figure 4.1 (right), if we add these subsequent splinter groups, the sample size grows to 349 organizations and the rate of splintering increases to 33.24%.[h]

---

[f] In cases where sources disagreed, I would search for an additional source to adjudicate between the two.

[g] Even with this extensive research, I could not find information in about 3.5% of the 85 cases of known group splintering from then random sample of 300. For five others, it was clear that a split was multidimensional, but sources disagreed on the exact reasons. These are coded as multidimensional but the specific disagreements are coded as missing. The full, extended list of organizations that includes splinters emerging from the random sample of 300 (which are *not* analyzed in the quantitative analyses) are listed in the Appendix.

[h] In the statistical tests I restrict my analyses to the original sample of 300 groups to preserve the benefits of random selection. I present the full list of groups I statistically evaluate in the Appendix.

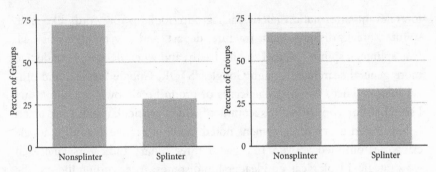

Fig. 4.1 The proportion of splinter and nonsplinter groups

Second, are there differences in the longevity between armed groups that do and do not splinter? If those that splinter tend to last significantly shorter than those that do not, breaking apart armed groups could very well form part of a successful counterinsurgency or counterterrorism strategy. The data suggest that groups that splinter tend to last slightly *longer* than those that never split. Restricting the analysis to groups that are no longer operating in the final year of the data set (since we cannot know their fate) reveals that groups that never splintered have a mean age of 5.6 years, while groups that *did* splinter lasted an average of 14.9 years. This is a statistically significant difference of nearly 9 years. Though interesting, this is not entirely surprising: factionalism tends to increase as groups age, and older organizations simply have a longer period during which fragmentation is possible.

Perhaps armed groups that do and do not splinter are different along other metrics. Few differences appear when one factor is taken into account: group age. For instance, groups that splintered launch on average 48.3 attacks with a standard deviation of 20.014, while those that never splinter launched 187.33 with a standard deviation of 61.101 and the difference is therefore significant at the .001 level. However, if we standardize the number of attacks by group age, the difference becomes statistically insignificant (p = .338). The same is true for total fatalities and casualties as well. These findings are displayed in Table 4.1, and they add further evidence to the conclusion that there are no meaningful differences between groups that do and do not break apart.

Third, do groups that *form* by splintering, regardless of their motivations, last long longer than nonsplinters? One could imagine that splinters gain some operational advantages as they are established with a core of trained fighters. Novice fighters—whether in armed groups or national militaries— are more likely to defect or disobey orders when facing live fire for the first

Table 4.1 Differences across organizations that do and do not splinter

Category	Splintered	Never Splintered	Significance
Mean Age	15.586 (.601)	14.933 (1.541)	.000
Mean Fatalities	144.931 (55.010)	383.766 (184.980)	.096
Mean Standardized Fatalities	20.465 (5.545)	21.050 (6.736)	.950
Mean Casualties	285.735 (92.594)	808.55 (375.265)	.054
Mean Standardized Casualties	50.373 (13.855)	43.164 (13.28)	.751
Mean Attacks	48.310 (20.014)	187.333 (61.101)	.005
Mean Standardized Attacks	6.835 (1.875)	9.882 (2.171)	.338

Metrics standardized by group age. Standard errors in parentheses. Groups active in final year of data set (2012) omitted.

time.[12] Splinters might be immune to this problem since their members previously faced combat. I find that splinters and nonsplinters endure at roughly equal rates. As measured by the number of years from their first known attack to their last using data from the Global Terrorism Database, the mean duration for nonsplinter organizations is 9.9 years while for splinters it is 8.2 years. T-tests find that the differences are statistically insignificant. States would therefore seem to derive little benefit from counterterrorism policies that break apart organizations since the resulting splinters are neither systematically more nor less durable.[i]

Fourth, are certain types of armed groups more likely to splinter than others? Perhaps religiously inspired organizations, for instance, are less prone to fracturing owing to the non-pecuniary benefits that motivate their fighters and demand their cooperation.[13] In terms of basic identity characteristics from the Terrorist Organizational Profiles data set, there are no meaningful differences between groups that do and not fracture. Patterns of fragmentation are roughly consistent across communist-socialist, leftist, nationalist-separatist, racist, and religiously motivated groups.[j] These results are displayed in Table 4.2.

Fifth, is splintering consistent over time? If organizational fractures were common in previous decades while waning in recent years, one might question the significance of this research to contemporary conflicts. This is not the case. Figure 4.2 plots the formation of splinter groups over time among

---

[i] If my theory is correct, this is what I would expect to find. Differences should only emerge when disaggregating splinters by their reasons for forming.

[j] Using chi-squared tests to compare these binary indicators of group identity, not a single one registers any meaningful difference across splinter and nonsplinter organizations.

Table 4.2 Frequency of splintering across common group identities

Identity	# Never Splintered	# Splintered	Chi^2
Communist-Socialist	39	15	.682
Leftist	20	6	.751
Nationalist-Separatist	75	32	.170
Racist	6	1	.485
Religious	47	20	.350
Total Observations	174	60	—

Groups active in final year of data set (2012) omitted.

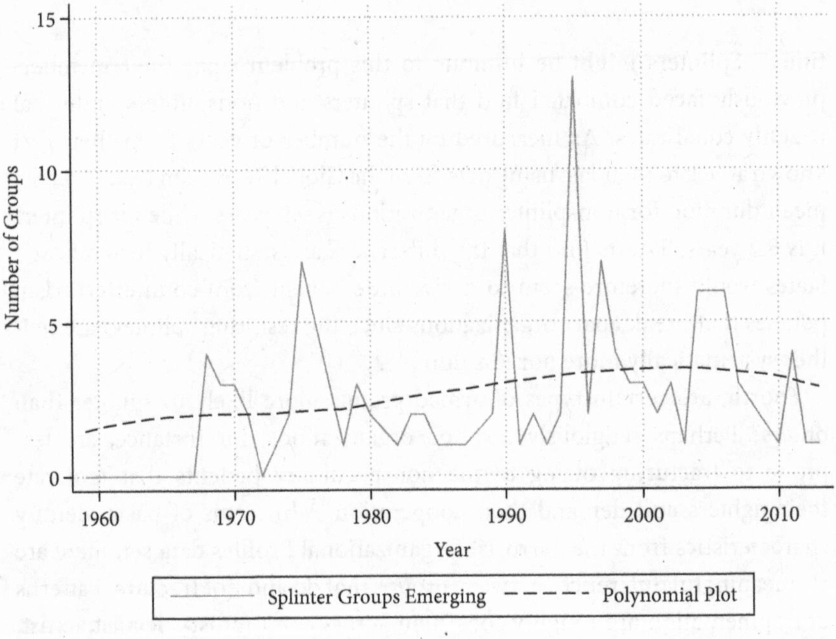

Fig. 4.2 Splinter formation over time

the random sample of 300 organizations. The blue line is a count of the number of new splinters identified each year while the red line is a polynomial regression line plotted to help visualize trends over time. Since the late 1960s, when the data begins, organizational splintering has been relatively constant.[k] There are, however, several points where splintering occurred at an increased rate: the late 1970s, the mid-1990s, and the mid-2000s. This

---

[k] There is only a single year in which no new splinter group is recorded: 1972. This would likely be resolved with a larger sample.

is not all that surprising. In the 1970s, Marxist and other leftist ideologies were common ideological flashpoints among armed groups. For instance, there was an intense ideological struggle within the People's Mujahedin of Iran (MEK) that culminated, in 1973, with MEK leaders expelling non-Marxist members. A group pamphlet even declared "that after ten years of secret existence, four years of armed struggle, and two years of intense ideological rethinking, they had reached the conclusion that Marxism, not Islam, was the true revolutionary philosophy."[14] Increased splintering in the 1990s could be linked to the collapse of the Soviet Union, the end of the Cold War, and the drying up of support for armed groups from the two superpowers.[15] This financial and material shock may have exacerbated intragroup tensions. Splintering in the mid-2000s could be linked to US military intervention in the Middle East and significantly more ambitious, coordinated, and effective counterterrorism campaigns as part of the Global War on Terror.

Finally, this new data reveals useful information about why armed groups break apart. When it comes to the most basic distinction—if splits are unidimensional or multidimensional—most (62%) organizational fractures are driven by a single disagreement. This could mean that individuals who consider breaking away from an organization do not wait until a multitude of disagreements exist; rather, they can be sufficiently motivated by even one dispute. When subgroups view them as matters of life and death, then a sole disagreement over strategy or leadership may be enough to break the organization apart.

In terms of their specific disagreements, the data reveal that strategy is involved in just over 60% of all fractures—whether unidimensional or multidimensional—followed by leadership disputes at 35%, and ideological

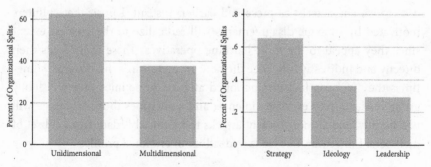

Fig. 4.3 Relative frequencies of organizational splits

disagreements at 30%.[1] When armed groups break apart, it is therefore most commonly due to disagreements over the use of force. This is not all that surprising since violent strategies are contentious, impactful, and also highly dynamic. They must be responsive to changing organizational priorities and resources, to competition and cooperation from other groups, and even the government's repressive and conciliatory actions. In effect, decisions about the use of force are constantly taking place throughout an armed campaign, presenting ample opportunities for disagreement.

Taken together, these basic analyses confirm that the phenomenon of organizational splintering is neither unique to a particular subset of militant groups nor to a particular historical period. Rather, internal schisms affect armed groups consistently over time and regardless of their ideological leanings or operational behavior. They also reveal that armed groups are commonly divided over issues of strategy, but ideology and leadership disputes each feature into about a third of organizational ruptures.

The Radicalization of Splinter Groups

I now evaluate why some splinter organizations radicalize to a greater extent than others. This has important policy implications. If states can understand which groups will develop into increasingly capable threats then they can devote more resources and enact policies earlier to defeat them. By understanding why some organizational fractures produce radicalized offshoots, states could also to take actions, or avoid actions, that make these splinters likely.

I argue that splinter groups are neither inherently more nor less violent than any other organization. This notion receives support from the basic descriptive characteristics presented earlier. Instead, I posit that splinters motivated by strategic disagreements will radicalize to the greatest extent since they appeal to the most hardline operatives. These individuals then directly and indirectly influence the new group towards the extreme: sometimes they will conduct unsanctioned attacks; sometimes they will lobby group leadership to escalate violence; and sometimes those in charge will unilaterally launch more violent attacks to placate this dangerous subset of members. Either way, this leads to deadlier operational profiles.

---

[1] These sum to over 100% since some ruptures have multiple causes.

The statistical analyses presented in the following pages largely confirm my expectations. Disagreements over strategy produce breakaway groups that are meticulous purveyors of violence. Their typical attack kills more people than those committed by any other group, whether splinter or nonsplinter. While they do not commit more attacks or kill more people in a given year, their individual operations are highly effective and highly lethal. This is a dangerous pattern. I also find that they exhibit this behavior almost immediately upon forming and it then carries throughout their entire lifespans. Organizational fractures driven by strategic disputes thus portend worrying consequences.

Measuring and Modeling the Radicalization of Armed Groups

To begin with, a term like radicalization can mean many things.[16] Here, I conceptualize it as tactics, strategies, and behaviors that are increasingly lethal and extreme. More specifically, I analyze four metrics of radicalization, and increases in each should correlate with increasingly radicalized behavior. They are attack frequency, attack lethality, cumulatively (yearly) lethality, and the employment of suicide bombs.

First, radicalization can be examined in terms of attack frequency. Groups that launch more attacks are increasingly radicalized as they ramp up the production of violence, causing more destruction and more fear. Second, I examine the total number of fatalities armed groups cause in a given year. Groups responsible for increased numbers of civilian fatalities can be considered more radical than others as they are able, and willing, to cause more destruction and loss of human life. Third, I examine average attack lethality, measured as the total number of fatalities divided by the total number of attacks in a given year. This sheds light on the typical deadliness

Table 4.3 Measuring and modeling the radicalization of armed groups

Indicator	Statistical Model
Attack Frequency	Negative Binomial Regression
Yearly Lethality	Negative Binomial Regression
Mean Attack Lethality	Negative Binomial Regression
Suicide Bombing	Logistic Regression

of a group's violent acts and it is a good indication of their relative embrace of violence.[m] Finally, I evaluate a group's tactical choices. Suicide bombings are some of the most destructive, violent acts that armed groups can perpetrate. According to the Global Terrorism Database, between 1970 and 2017 the average suicide attack caused nearly 31.14 casualties (individuals killed *and* wounded), whereas the average non-suicide attack caused 4.29. These acts are therefore radical from the standpoint of killing capacity, but the calculated suicide component—whereby the perpetrator intentionally perishes as part of the operation—is itself a radical act. I measure each of these variables per year for all groups in my data set using information from the Global Terrorism Database.

In terms of statistical models, I analyze the frequency of attacks, the average lethality of attacks, and the number of fatalities with negative binomial regressions—a model that is appropriate for non-negative count variables that follow an unknown distribution.[17] In the analysis of suicide bombing, with a binary dependent variable, I employ logistic regressions. For both, I cluster standard errors by group to account for unexplained differences across organizations.

My primary explanatory variable—what I posit is partly driving rates of radicalization—is how splinter groups form. I begin by evaluating if there is a distinction between splinter and nonsplinter groups. Then, I assess differences between nonsplinter groups, splinters motivated by strategic disagreements, and other splinters. To do so I create a categorical variable that takes on a value of zero for nonsplinters; one for strategically motivated splinters (with strategy being their sole disagreement or as part of a multidimensional rupture), and two for other splinters. Finally, I evaluate whether unidimensional schisms centered on strategy have distinct effects. Since the resulting breakaway groups have a singular focus on violence, they may be especially likely to radicalize. To evaluate, I create a categorical variable coded as zero for nonsplinters, one for splinters motivated solely by ideology, two for strategy, three for leadership, and four for multidimensional disagreements.

---

[m] It is worth noting that these attack statistics might also increase when armed groups become more efficient, and not necessarily more radicalized. This is certainly true and there is an important element of learning that occurs within armed groups. I control for the influence of learning and assistance in two ways: I include covariates that capture group age (assuming that older groups have more experience) and external alliances; and I perform a number of analyses on group behavior in their first year of recorded activity when differences in experience, alliances, and other factors are minimized.

While my theory predicts that breakaway groups emerging from strategic disputes are the most likely to radicalize, other factors undoubtedly shape their tactics and strategies. To account for these, and to ensure that the statistical models are finding variation that is driven by how splinters form, I implement two strategies. First, I include country and year fixed effects in most statistical models. This accounts for all unchanging dynamics that vary across countries and over time, from a culture's willingness to support violence to a state's proportion of rugged and mountainous terrain to the changing propensity for foreign states to intervene in subnational affairs. This is a robust strategy to account for factors that I cannot directly measure or even observe.

Second, I include a host of variables with theoretical links to rates of radicalization. Among existing explanations for why groups radicalize, chief among them is intergroup competition. Multiple groups in the same environment (typically conceptualized at the state level) will prompt groups to up the ante of violence and conduct increasingly violent, destructive acts in an attempt to consolidate control and prove their dedication over rivals.[18] And as Bloom notes, "The more spectacular and daring the attacks, the more the insurgent organization is able to reap a public relations advantage over its rivals and/or enemies."[19] To proxy for outbidding, I use the Global Terrorism Database to generate a count of the number of active organizations in a given country-year.[n] With more groups, more outbidding and greater radicalization is expected.

Particular properties and characteristics of individual armed groups are also linked to rates of radicalization. Scholars find that tactical knowledge is important. When organizations possess increasing numbers of inter-organizational alliances they can share their information and skills, contributing to the diffusion of tactical capabilities across organizational lines.[20] Researchers also find that particular types of organizations are more lethal than others, and this is especially true for religious groups.[21] Identity matters for many reasons: it can shape tactical preferences, social networks, the motivation of recruits, and so on.[22] Lastly, both a group's age and its transnational nature should also be linked to radicalization. Older groups might be more knowledgeable and more effective, leading to deadlier operations.

---

[n] I first removed attacks by unknown actors and also from perpetrators that were not part of an actual group (e.g., nondescript attacks by "students" or "protesters"). I then summed the number of groups in a country-year that committed at least one attack and subtracted one to remove the group being analyzed.

As for being transnational, groups operating across borders could access resources and safe havens that yield operational capacity.

Data on group identities and alliances is taken from the RAND-MIPT project on Terrorist Organizational Profiles (TOPs). To account for the influence of group identity on rates of radicalization I include dummy variables indicating nationalist-separatist, communist-socialist, religious, and leftist ideologies. Group alliances are counted as the total number of other organizations that a particular group was known to cooperate with.[o] I also use attack statistics from the GTD to construct a dummy variable indicating transnational capabilities (whether a group launched an attack in at least two separate countries in a given year), and I calculate a group's age as the number of years since its first known attack.

State characteristics are undoubtedly important as well. There is an extensive body of work on why and how the basic characteristics of autocratic and democratic governments influence the dynamics of subnational violence. Although the specific actions that governments take to repress and to restrict militant networks undoubtedly influences their strategic calculus,[23] these studies focus more on the innate characteristics of particular regime types. For instance, democracies can encourage violence by prompting marginalized groups to seek non-political means to achieve their goals,[24] by restricting the range of potential defensive measures,[25] and by nature of their popular decision-making, democratic citizens are convenient targets.[26] Finally, state capability may be linked to the ability to combat armed groups and reduce violence.[27]

To account for the effects of regime type I use the Polity IV data set[28] to construct dichotomous indicators for democracy and autocracy. Polity scores range from −10, most autocratic, to 10, most democratic. I code democracies as 7 and above, and autocracies as −7 and below. I also include a measure of regime durability from the Polity project that counts the number of years that the current regime has existed. Here, the logic is that older and more established regimes with more developed institutions might be better able to combat violent extremists and even less likely to witness violence in the first place. Other state characteristics that might be significant include wealth and regime durability. GDP per capita, for instance, might interact with radicalization in three ways: funding for militants might be

---

[o] Both the measures for group ideologies and alliances are measured cross-sectionally and do not change over time.

greater in wealthier countries, making it easy for groups to purchase weapons and other supplies; wealthier countries might have more established and better equipped police forces which can more easily confront violent organizations; finally, if one believes that militant violence is a function of economic grievances, then wealthier nations with higher GDPs per capita might experience less severe subnational violence. Data for GDP per capita and population come from the Penn World Tables.

Statistical Results I: Initial Behavior

I start by examining the characteristics of organizational violence—the number of attacks, fatalities, and the average lethality—in the first year of an armed group's recorded activity. I only include year fixed effects because of the limited number of observations.[p] Recall that these results are derived from negative binomial regressions. So, coefficient estimates above one indicate higher rates of radicalization, and below one lower rates of radicalization.

This first year analysis provides a crucial view into a group's earliest behavior. At this point, armed groups are relatively uninfluenced by intergroup alliances (which are unlikely to have formed yet), competition, and other social, political, and organizational changes that develop over time. Rather, they are typically drawing upon innate organizational capabilities that are present from the start. I further expect differences between splinters and non-splinters to emerge at this point because members of the former already have experience in clandestine activities. Their members retain important skills and technical know-how that should enable splinters to more effectively hit the ground running.[q]

---

[p] This is only intended to offer a preliminary glimpse into the results while more rigorous estimates are presented in the following section.

[q] Of course, this dichotomy—with splinters forming with experienced members and nonsplinters without—is certainly not universally true. Individuals with experience in insurgent or terrorist violence can also form an entirely new organization or even join a preexisting group. This is one reason to explain the considerable success and ability of groups like Al-Qaeda and the Taliban: both organizations benefited and even capitalized upon the experience and training their members received from past conflicts. Prominently, former members of the Afghan Mujahideen joined both of these groups following the Soviet withdrawal from Afghanistan in 1989, bringing with them tactical and operational expertise that positively contributed to the capabilities of their new organization. Nonetheless, while nonsplinters *sometimes* form with experienced members, splinters always do. However, I still expect splinters emerging from strategic disputes to be the most highly radicalized even in this first year. While all splinters should have a core of knowledgeable members, as just

Table 4.4 Splinter and nonsplinter violence in year one

	Yearly Attacks	Yearly Fatalities	Yearly Average Lethality
Splinter Group	1.033	1.974**	1.428
	(0.149)	(0.636)	(0.417)
Controls	No	No	No
Fixed Effects	Year	Year	Year
Observations	287	287	287

Standard errors in parentheses (clustered by group).
Results reported as incident rate ratios. * $p<.1$, ** $p<.05$, *** $p<.01$

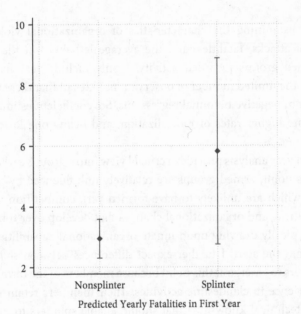

Fig. 4.4 Predicted fatalities in year one for splinter and nonsplinter groups

The results in Table 4.4 show that splinter groups are responsible for almost 97% more fatalities than nonsplinters in their first recorded year of activity, but they conduct similar numbers of attacks and have similar per-attack lethalities. To illustrate, Figure 4.4 plots the predicted number of first-year fatalities, and we can see how the number is much higher for breakaway

discussed, those motivated by strategic differences should be the most likely to put that capability to use. Intention and motivation should be not be ignored when explaining the behavior of armed groups.

Table 4.5 Strategic splinter violence in year one

	Yearly Attacks	Yearly Fatalities	Yearly Average Lethality
Splinter, Strategy	0.990	2.763***	2.242**
Involved	(0.165)	(1.004)	(0.767)
Splinter, No Strategy	1.125	0.825	0.501*
	(0.289)	(0.419)	(0.194)
Controls	No	No	No
Fixed Effects	Year	Year	Year
Observations	287	287	287

Standard errors in parentheses (clustered by group).
* *p<.1*, ** *p<.05*, *** *p<.01*

groups. This is evidence that splinters, regardless of why they emerge, are especially deadly organizations in their first year of operation. This is perhaps testament to the innate experience they bring to bear.

Next, I further evaluate first-year differences between nonsplinter groups, splinters that form over strategic disagreements (as their sole motivation or otherwise), and splinters that form over other issues. These are displayed in Table 4.5. In line with my expectations, I find that strategic disputes produce the most radicalized armed groups of all: in their first year, they generate 176% more fatalities and 124% more fatalities in each individual attack. These figures are significantly higher than those found in the previous analysis for all splinters, and they underscore the danger posed by this subset of groups. In contrast, splinters forming over other issues are usually *less* radicalized in their first year. Those emerging from disagreements over ideology or leadership absent any strategic component exhibit a 50% lower per-attack lethality. This shows how the grievances leading splinters to emerge are linked to meaningful differences even in their short-term trajectories.

Finally, I evaluate differences between splinter groups forming unidimensionally and multidimensionally in Table 4.6. It could be that splinters are most likely to radicalize when solely motivated by strategic grievances, leading to a particularly homogeneous core of disaffected hardliners. Splinters forming multidimensionally over strategy and other issues may not experience the same, consistent pressure to ramp up violence. The results, graphically depicted in Figure 4.5, show this to be true: unidimensional splits over strategy produce breakaway groups that generate 212% more fatalities and exhibit 124% higher attack lethality in their first year (than similar nonsplinter groups). This is notably higher than the figures identified above

Table 4.6 Disaggregating splinter violence in year one

	Yearly Attacks	Yearly Fatalities	Yearly Average Lethality
Strategic Splinter	1.004	3.127***	2.243**
	(0.195)	(1.352)	(0.915)
Ideological Splinter	3.596**	4.034**	1.299
	(1.821)	(2.460)	(0.769)
Leadership Splinter	0.618	0.084***	0.145***
	(0.329)	(0.044)	(0.069)
Multidimensional	0.718*	0.694	0.677
Splinter	(0.123)	(0.289)	(0.299)
Controls	No	No	No
Fixed Effects	Year	Year	Year
Observations	286	286	286

Standard errors in parentheses (clustered by group).
Results reported as incident rate ratios. *$p<.1$, ** $p<.05$, *** $p<.01$

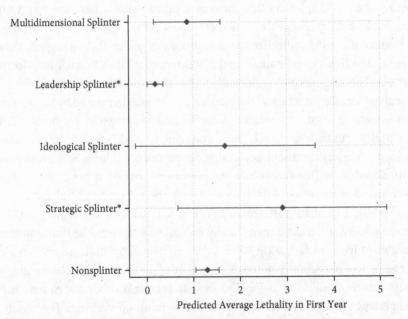

Fig. 4.5 Predicted average lethality in year one for different splinter groups
Significant estimates (from original model estimation, relative to nonsplinter groups) are marked with an asterisk

where strategy was only one reason among several causing groups to break with their parents.

Interestingly, the results show that splinters emerging from ideological disagreements are also quite dangerous. They launch more attacks and cause more fatalities than nonsplinter groups, but they exhibit a similar per-attack lethality. If these groups are seeking to establish their ideological credentials and dedication in the wake of an internal dispute, they may seek to prove themselves with violence.[29] In fact, it is plausible that such competition would be most severe in a group's first year when the need to establish their identity is most pressing. Conversely, I find that splinters emerging from multidimensional and leadership disputes exhibit slightly lower rates of radicalization in their first year. Leadership disputes generate groups with lower yearly fatality and average lethality rates, and for multidimensional splits, the new groups also conduct fewer attacks. It could be that the weaker internal cohesion of these organizations negatively affects their operational capability. Groups beset by internal feuding and low cooperation might find it difficult to plan and then launch successful attacks. Consequently, these groups might not necessarily choose to be less radical, but their internal preference divergence prohibits them from sustaining violence.

Statistical Results II: Sustained Behavior

I now assess rates of radicalization throughout the entire lifespan of armed groups. I study their behavior in every recorded year activity, and the analyses include the full list of covariates and fixed effects described earlier. As a reminder, these covariates account for factors beyond how splinters form that should affect how they operate.

As before, I begin by assessing whether splinter groups operate distinctly from non-splinters. These results are presented in Table 4.7. Only one significant difference emerges: groups that form by splintering tend to conduct *fewer* attacks in a given year, as indicated by the statistically significant coefficient estimate below one. They are responsible for around 31% fewer attacks than nonsplinter groups in a given year, holding all other factors constant. Interestingly, when analyzing group behavior in the first year alone, I find that the number of attacks is largely equal. This suggests that many of these groups become slightly less active over time. No other differences emerge.

Table 4.7 Splinter and nonsplinter violence

	Yearly Attacks	Yearly Fatalities	Yearly Average Lethality
Splinter Group	0.694**	0.778	1.177
	(0.121)	(0.149)	(0.139)
Controls	Yes	Yes	Yes
Fixed Effects	Country, Year	Country, Year	Country, Year
Observations	1877	1877	1877

Standard errors in parentheses (clustered by group).
* p<.1, ** p<.05, *** p<.01

Table 4.8 Strategic splinter violence

	Yearly Attacks	Yearly Fatalities	Yearly Average Lethality
Splinter, Strategy	0.735	0.849	1.288**
Involved	(0.149)	(0.185)	(0.166)
Splinter, No Strategy	0.564**	0.536*	0.838
	(0.127)	(0.176)	(0.179)
Controls	Yes	Yes	Yes
Fixed Effects	Country, Year	Country, Year	Country, Year
Observations	1877	1877	1877

Standard errors in parentheses (clustered by group).
* p<.1, ** p<.05, *** p<.01

Next, in Table 4.8, I assess operational differences between splinters that form over strategic disputes and those that do not. When strategy motivates their departure, breakaway groups exhibit a 28% higher average attack lethality that persists throughout their entire existence. In effect, the typical nonsplinter group is predicted to have a typical per-attack lethality of 3.4, but it is 4.4 for splinters motivated by strategy. These same groups are not expected to conduct more or fewer attacks, or to kill more people overall. Instead, they are just more efficient and seemingly more calculated with their violence, killing more people with every single operation.

On the other hand, when strategy is not an issue and hardliners are not enticed, breakaway groups moderate their behavior. They conduct nearly 50% fewer attacks and kill nearly 50% fewer people, but maintain a comparable lethality to nonsplinter groups. This is the inverse of what I find among their first year behavior, and it underscores how much organizations can evolve from their first year onwards. It also shows how strongly membership composition is linked to the behavior of armed groups.

Finally, in Table 4.9 I evaluate how groups behave when forming unidimensionally and multidimensionally. As noted earlier, I expect unidimensional splits over strategy to produce the most radicalized offshoots since this should generate a particularly homogeneous core of disaffected hardliners. The results again support this intuition: splinter groups forming unidimensionally over strategy exhibit a 50% higher per-attack lethality across their entire existence. The predicted average lethality is depicted in Figure 4.6. The fact that such groups are killing many more people in every operation is no small matter: the vast majority of terrorist attacks do not produce any casualties whatsoever, and even small attacks depend upon significant planning and logistic networks. In effect, these breakaway groups are highly determined to commit deadly acts of violence, and they are highly successful at doing so.

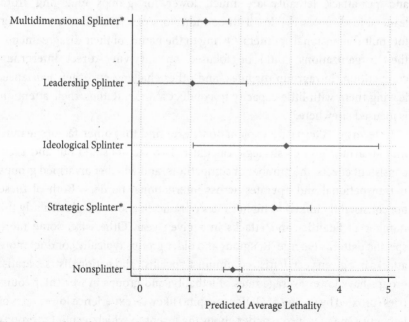

Fig. 4.6 Predicted average lethality for different splinter groups

Significant estimates (from original model estimation, relative to nonsplinter groups) are marked with an asterisk

Table 4.9 Disaggregating splinter violence

	Yearly Attacks	Yearly Fatalities	Yearly Average Lethality
Strategic Splinter	0.809	0.992	1.440**
	(0.155)	(0.221)	(0.214)
Ideological Splinter	0.887	1.233	1.561
	(0.287)	(0.517)	(0.515)
Leadership Splinter	0.262***	0.150***	0.554
	(0.090)	(0.098)	(0.297)
Multidimensional	0.499**	0.372***	0.695*
Splinter	(0.165)	(0.143)	(0.138)
Controls	Yes	Yes	Yes
Fixed Effects	Country, Year	Country, Year	Country, Year
Observations	1876	1876	1876

Standard errors in parentheses (clustered by group).
* p<.1, ** p<.05, *** p<.01

When it comes to other disputes, I find that ideological disagreements generate splinters that are comparable to nonsplinters. However, fatalities and per-attack lethality are much lower for groups emerging from leadership disputes, while all three indicators of radicalization are lower for multidimensional splinters. Owing to the nature of their disagreements, these organizations could be focused on surviving direct intergroup competition, intragroup disputes, and other challenges to their existence, leaving them with little capacity to conduct attacks. Rather, their attention is focused elsewhere.

In terms of alternative explanations, I do find that other factors feature into an armed group's strategic calculus.[r] Two factors stand out and exert consistent effects: the number of competitors, and whether an armed groups is transnational and operates across international borders. Both of these are consistently linked to higher rates of radicalization and particularly the number of fatalities and attacks in a given year. Otherwise, some more specific patterns emerge. Religious and older groups typically conduct more attacks in a given year; leftist, communist-socialist, and nationalist-separatist groups have lower average rates of lethality; and groups in wealthier countries (proxied by the log of GDP per capita) likewise experience lower rates of lethality. One may also wonder about the extent to which a splinter group's rate of radicalization is linked to the behavior of its parent. Maybe more radicalized parents produce more radicalized splinters. I test this proposition

[r] The full models with controls are available in the Appendix.

in the Appendix and I find no support for it. Analyzing only splinter groups, I find a weak negative connection between the parent's average lethality and that of a splinter. The number of attacks and fatalities appear unrelated, while the motivations underlying a splinter's exit continue to exert one of the most significant effects of the entire model.

Taken together, these analyses once again reveal that how armed groups form is closely connected to their operational profiles. But there is now confirmation that these effects are not temporary: they not only shape group behavior in their formative years, but they are statistically detectable throughout the entire lifespans as well. Most prominently, the data reveal that splinter groups motivated by strategy are not especially active or lethal on a yearly basis, but every one of their attacks is especially deadly. The emergence of these groups is therefore a troubling development.

Statistical Results III: Suicide Bombing

Now I evaluate whether certain types of breakaway groups are more or less likely to employ suicide bombing in a given year. As an indicator of radicalization, it is plausible that splinters motivated by strategic disagreements will utilize human bombs to a greater extent than others. The statistical results, which mirror the model specifications from before, are presented in Table 4.10. Since the estimates are derived from logistic regressions, coefficients above one imply a higher likelihood of suicide bomb usage, and below one a lower likelihood.

In contrast to my expectations, breakaway groups motivated by ideological disputes are most likely to use suicide bombs. To begin with, splinter groups as a whole are neither more nor less inclined to use them. When it comes to organizational splits involving strategic disagreements, these groups are just over 60% *less* likely to use suicide bombs. Interestingly, it is unidimensional splits over ideology that create groups with the highest proclivity for this particular tactic. They are 300% more likely to launch a suicide bomb in a given year than the typical nonsplinter group (while controlling for other covariates, and both country and year fixed effects). This is a large, substantive increase.

Obviously, this goes against my theoretical intuition. Hardliners do *not* seem keen to utilize suicide bombs, and this deviates from the patterns of radicalization identified so far. Why might this be the case?

Table 4.10 The use of suicide bombing

	Model 1	Model 2	Model 3
Splinter Group	0.603		
	(0.254)		
Splinter, Strategy Involved		0.370**	
		(0.186)	
Splinter, No Strategy		1.793	
		(0.974)	
Strategic Splinter			0.462
			(0.237)
Ideological Splinter			4.576**
			(2.943)
Leadership Splinter			1.000
			(.)
Multidimensional Splinter			0.716
			(0.961)
Controls	Yes	Yes	Yes
Fixed Effects	Country, Year	Country, Year	Country, Year
Observations	954	954	941

Standard errors in parentheses (clustered by group).
* $p<.1$, ** $p<.05$, *** $p<.01$

First, among armed groups, suicide bombing is not solely relegated to the most radicalized organizations. Among the top ten deadliest groups listed in the GTD (by total number of fatalities caused), four never used it. This includes the Shining Path in Peru, Farabundo Marti National Liberation Front, the Nicaraguan Democratic Force, and the Armed Revolutionary Forces of Colombia. Instead, it is most common to a specific subset of groups in the top ten: those motivated by religion, mostly located in the Middle East and North Africa, and with connections to other groups with relevant experience. Ideology plays such an important role to the development and use of suicide bombs since these operations require unwavering and unparalleled dedication by operatives, and for groups to accept the potential reputational costs that go along with them. Taken together, suicide bombing is not necessarily the logical conclusion for radicalized members of every armed group, but only for some of them.

Second, since ideology is so central to the use of suicide bombing, it is understandable that groups breaking away over ideological differences may be more inclined towards this particular method of attack. If these splits operate similarly to disputes over strategy, it could be that ideological disputes are frequently instigated by ideological "hardliners" who hold more extreme, rigid, and devout views. Their ideological extremism may explain why they gravitate towards these specific operations. To be certain, however, we need more research that looks into the dynamics of these particular fractures more closely.

The Longevity of Splinter Groups

I have so far showed that patterns of organizational breakdown are linked to the radicalization of breakaway groups, and here I assess whether these patterns extend to their longevity. As before, this is an important puzzle: some splinters are able to withstand government pressure and threaten public safety for years while others quickly collapse. Understanding this variation can help policymakers design initiatives that contribute to splinter groups' decline, to focus these policies on the groups most likely to pose enduring challenges, and to avoid policies that might create durable splinters in the first place. This information can also help identify the types of organizations that are likely to disintegrate on their own, absent state intervention.

My theory implies that splinter groups will survive longer when forming unidimensionally over single, shared grievances that unite members and build cohesion. They should be less stable when forming multidimensionally over multiple grievances that attract disparate members with competing visions. Statistical evidence confirms this intuition. Controlling for a variety of confounding factors, I find that unidimensional fractures produce groups that are 10% more likely to survive past their first year and 25% more likely to survive past their tenth year than those emerging from multidimensional fractures. This supports my theory that niche size and internal preference consistency, as proxied by the unidimensional-multidimensional nature of group schisms, is critical to organizational longevity.

Measuring and Modeling the Survival of Armed Groups

I conceptualize a militant group's survival or its longevity (which I use interchangeably) as the duration of its violent activities. In other words, it is the difference in time between when a group begins violent activities and when it commits its last attack.[30] To construct this variable, I begin with the data set of 300 randomly selected militant organizations I presented earlier. Using individual attack statistics, I then calculate the age for each organization in years. Groups begin with their first known attack and end with their last.

To statistically model how long armed groups endure, I employ the Cox Proportional Hazards Model, a form of survival analysis.[s] The Cox model estimates the survival function that provides information about the likelihood of failure (groups ending) in a given year. The underlying idea is that different factors or covariates shift the hazard up or down, increasing or decreasing the chance of failure at a given point in time. The model is proportional in the sense that the estimated effects are assumed to be consistent over time.[t] For instance, it assumes that the effects of operating under an autocratic regime are the same for organizations in their first year of activity and their tenth year of activity.[31]

In line with my theory, the primary independent variable that should partially explain rates of survival is the unidimensional or multidimensional nature of an organizational split. This is coded using a categorical variable where zero refers to a nonsplinter group, one to a unidimensional splinter, and two to a multidimensional splinter. In subsequent analyses, I also examine the specific motivations underlying unidimensional ruptures.

Of course, other factors undoubtedly affect how long armed groups survive, and I include many of these in my analyses. An organization's identity and goals might affect its longevity by providing a means to unite members while also enabling group leaders to reach a particular pool of recruits. The group identities I account for are: nationalist-separatist, communist-socialist, religious, and leftist. In this system, groups can belong to several different identities at once. An organization can be both

---

[s] The Cox model is designed for count data and it can incorporate time-varying covariates and information on censored observations, making it ideal for this specific task. It is a semi-parametric estimator that makes fewer assumptions than other parametric models about the underlying data-generating process and the functional form of the hazard function. If we have information about the hazard function and underlying data-generating process, parametric survival analysis is more efficient. However, failing to correctly specify the distribution of hazard function can produce incorrect and misleading results. For instance, if one could assume that the underlying data-generation process followed a Weibull or Poisson distribution, then a parametric estimator would be a better choice than a semi-parametric estimator. With the Cox model, however, scholars need not make assumptions about how the hazard function is distributed. It can more flexibly model data about which little is known.

[t] Proportionality is a critical assumption underlying this model. In the Appendix, I evaluate whether my main independent variables do in fact generate proportional effects, and the results confirm this assumption. This supports using this particular model specification.

nationalist-separatist and communist-socialist, for example. With regard to armed groups in particular, religion is especially important.[32] Groups can potentially draw upon their shared faith to enhance cooperation and unify the group in a way that bolsters their survivability.[33] I also account for particular state characteristics. Existing literature finds strong theoretical and empirical links between regime type and group proliferation,[34] as well as regime type and states' capacity to fight militants.[35] Young and Dugan argue that because "democratic societies offer institutional recourse for aggrieved individuals, people have formal mechanisms for resolving their anger towards the state" and the resulting terror groups should be short-lived.[36] I therefore include dummy variables for autocracy and democracy. These come from Polity scores, and democracies are coded as 7 and above, and autocracies as −7 and below.[37] Relatedly, I include a dummy variable for regime durability measured as the total number of years since the most recent regime change.[38] Durability could be connected to organizational longevity through several mechanisms, but most significantly, more durable regimes might be better prepared to combat violent nonstate actors. I also include measures of a country's GDP per capita and its population.[39] GDP can influence militants' survival in several ways. Some scholars have argued that poor economic performance and low levels of development are correlated with militant activity and violent extremism.[40] These economic environments create grievances that motivate individuals to take up arms, make it easier to recruit new members who are facing diminished economic prospects, and lower the cost of joining the militant group when there are fewer opportunities for employment. These mechanisms are expected to work at the opposite end of the economic spectrum as well: greater economic prosperity should reduce popular grievances and subsequently the support for militant organizations. According to this logic, GDP per capita should be negatively correlated with the survival of armed groups.

Group competition is another potential explanation for how long groups endure. A more competitive environment could lead to an early demise, since resources and recruits are more difficult to come by. Direct intergroup competition might also weaken a fighter's commitment to a specific group if other groups offer similar opportunities. I approximate these dynamics with several measures: first, the number of active groups in a country-year; second, with a dichotomous measure of whether an armed group is the "top dog" (most active organization by number of in-country attacks according to the Global Terrorism Database) in a given year; and third, whether the

Cold War is taking place. The number of active militant groups in a given country-year is intended to capture the possibility of outbidding—a dynamic that is theorized to occur when multiple nonstate actors exist in the same environment. Since most resources are zero-sum, groups are inherently in competition with one another for recruits, popular support, et cetera.[41] The outbidding phenomenon is most commonly used to explain levels of violence in a particular country, but it might also "[dampen] group survival as other organisations drain the pool of potential recruits."[42] Young and Dugan also introduce the idea that an organization's status as "top dog" will correlate with that group's resiliency.[43] Using an analogy to firms in the marketplace, groups that already have a strong base of support are more resilient than others, making it less likely that they will die out in any particular year.

Of course, there are many factors that I cannot directly measure or even observe that could affect how long an armed group survives. To capture as many of these as possible, I include country and year fixed effects in every statistical model. This accounts for all unchanging dynamics that vary across countries and over time—from a culture's willingness to support violence, to a state's proportion of rugged and mountainous terrain, to the propensity for foreign states to intervene in subnational affairs. This is a robust strategy that can absorb much of the variation in longevity, but that can also rule out a host of competing explanations.

Statistical Results: Why Splinter Organizations Survive

I begin my analyses in Table 4.11. Here, I analyze the link between splinter formation and group longevity, but the statistical models only include country and year fixed effects without the control variables described above. In Table 4.12, the full controls are added. I do this to assess the extent to which my findings change with and without these additional covariates. If the results are relatively consistent, it is a sign of more robust statistical conclusions. If they change, it could signal that splinter dynamics are correlated with other variables in the model. For both analyses, Model One assesses the effect of groups forming as splinters or not, Model Two as splinters forming unidimensionally or multidimensionally, and Model Three as splinters forming according to different unidimensional disagreements.

Before proceeding, it is important to understand what the coefficients mean in these statistical models. Since the dependent variable is the time until group failure, the model estimates the *hazard* rate, or in other words, the likelihood that a group fails in a given year. A coefficient greater than one means that a given variable increases the hazard rate and a group is therefore *more* likely to fail. Coefficients below one lower the hazard rate and imply that a group is *less* likely to fail.

The findings in Model One, Table 4.11 (absent controls), reveal that there is no significant difference in how long splinter and nonsplinter groups survive. Their likeilihood of failure in any given year is largely consistent, and patterns of longevity are the same regardless of whether a group formed by splintering or not. This refutes the idea that splintering is inherently a useful strategy to defeat armed groups since those that emerge are no more or less likely to fail.

Next, the findings in Model Two support the idea that patterns of splinter formation are linked to rates of survival. Whereas splinter groups as a whole persist at rates similar to nonsplinters (Model One), I now find that groups forming unidimensionally are much more likely to persist. Groups forming multidimensionally, on the other hand, are much more likely to fail. Forming around a single, shared grievance therefore seems to give splinter groups a meaningful advantage.

Table 4.11 Cox proportional hazards model of organizational longevity without controls

	Model 1	Model 2	Model 3
Splinter	0.782		
	(0.145)		
Unidimensional Splinter		0.530***	
		(0.127)	
Multidimensional Splinter		1.687**	
		(0.397)	
Strategic Splinter			0.471***
			(0.124)
Ideological Splinter			0.418
			(0.316)
Leadership Splinter			0.852
			(0.351)
Multidimensional Splinter			1.680**
			(0.397)
Controls	No	No	No
Fixed Effects	Country, Year	Country, Year	Country, Year
Observations	2910	2910	2909

Standard errors in parentheses (clustered by group). Coefficients reported as hazard ratios.
* $p<.1$, ** $p<.05$, *** $p<.01$

Finally, Model 3 disaggregates unidimensional splinters according to the specific disagreements that motivated them to form: ideology, leadership, or strategy. This reveals that splinters forming unidimensionally over strategy are the most durable of all (relative to non-splinter groups, the reference category). While the estimates are for ideological and leadership splinters are in the anticipated direction—that is, lower hazard rates and greater longevity—they do not reach standard levels of statistical significance. In effect, this means that splinters forming over leadership and ideological debates survive and fail at rates similar to nonsplinter groups.[u] However, post-tests reveal that ideological and leadership splinters are significantly *more* durable than multidimensional splinters, supporting my theoretical intuition.

In Table 4.12 I replicate the results just presented while adding the full battery of covariates described earlier.[v] Recall that these variables aim to

Table 4.12 Cox proportional hazards model of organizational longevity with controls

	Model 1	Model 2	Model 3
Splinter	0.735		
	(0.141)		
Unidimensional Splinter		0.494***	
		(0.122)	
Multidimensional Splinter		1.582*	
		(0.386)	
Strategic Splinter			0.459***
			(0.122)
Ideological Splinter			0.326
			(0.247)
Leadership Splinter			0.699
			(0.314)
Multidimensional Splinter			1.584*
			(0.387)
Controls	Yes	Yes	Yes
Fixed Effects	Country, Year	Country, Year	Country, Year
Observations	2910	2910	2909

Standard errors in parentheses (clustered by group). Coefficients reported as hazard ratios.
* $p<.1$, ** $p<.05$, *** $p<.01$

[u] These same groups also cannot be distinguished from groups forming over strategy, either (as indicated by post-test chi^2 estimates). This implies that there are strong similarities among all groups produced by unidimensional fractures.

[v] And in the Appendix, I present additional tests using even more variables. For instance, I evaluate how state repression and an organization's lethality influences group longevity. The main findings do not change.

capture competing explanations for organizational longevity not picked up the country and year fixed effects from the previous set of analyses. These variables range from regime type to the group's age. I begin my analyses without them to assess how they influence my findings.

Overall, the results barely change. This suggests that the method of splinter formation exerts an effect in addition to these other variables now included. As before, the analyses reveal that splinter and nonsplinter groups survival and fail at similar rates; unidimensional splinters last longer than nonsplinters, while multidimensional splinters last shorter; and strategic splits produce the most resilient groups of all. Taken together, these findings support my theory and they underscore how the formation of armed groups is intimately linked to their future prospects, and specifically, their ability to survive.

To illustrate these effects, the survival patterns of nonsplinter groups and splinters forming unidimensionally and multidimensionally are plotted in Figure 4.7. Each line represents the odds that a group survives to the following year at a given age (indicated by the X axis). The solid black line is for the average nonsplinter group, the dashed line for unidimensional splinters, and the dotted line for multidimensional splinters. As one can see, the survival rate is significantly higher for breakaway groups forming

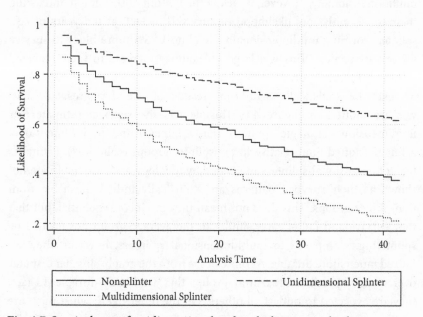

Fig. 4.7 Survival rate of unidimensional and multidimensional splinters

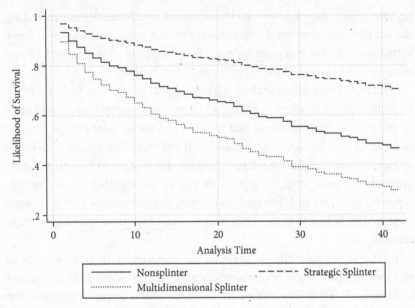

Fig. 4.8 Survival rate of strategic and multidimensional splinters

unidimensionally around a common disagreement, indicating that they are more durable and long-lasting threats. The line for splinters forming multidimensionally, however, is well at the bottom of this graph, indicating their consistently low likelihood of survival. In fact, at ten years of age, splinters forming unidimensionally are almost 25% more likely to survive for one more year. This is a large, substantive increase in their expected durability that comports with my theory.

Next, the statistically significant results of the disaggregated analysis are plotted in Figure 4.8. This depicts the survival estimates for unidimensional-strategic splinters (dashed lined) and multidimensional splinters (dotted line) relative to nonsplinter groups (solid line). Estimates for unidimensional splinters motivated by ideology and leadership are not shown as their survival patterns are statistically indistinguishable from nonsplinter groups. This does not mean they are inconsequential, but that they survive and fail at rates similar to the average group not formed by splintering. Compared to multidimensional splinters, however, they are indeed much more durable. As one can see from the graph, singular disputes over strategy produce breakaway groups that are highly durable. In fact, they are expected to outlast all others. On the other hand, when there are multiple disputes motivating groups to exit, their survival is uncertain. Their members are less unified and often disagree, and with lower rates of cooperation, they exhibit lower rates of survival. At ten years of age, their survival rate is 30% lower than that of groups forming unidimensionally.

In terms of alternative explanations, some factors do shape the longevity of splinter groups on top of their formative dynamics. Communist-socialist groups appear to be the most durable, with a lower odds of failure in any given year, while leftist groups are some of the least resilient. As for country characteristics, armed groups in larger states tend to be more ephemeral with a greater chance of failure. Other factors do not exert statistically significant effects. As I allude to elsewhere, these findings underscore that armed groups are simultaneously shaped by internal as well as external factors.

Overall, these results strongly support my theory. When it comes to why some splinter groups endure while others quickly fail, those emerging over single issue-areas and that attract a preference-aligned group of fighters who share a common vision for their organizational future maintain distinct advantages. As I show in my case study of Northern Ireland, this offers armed groups a signification advantage, making them more cohesive and more durable threats. But when they form multidimensionally, group cohesion suffers and they are prone to infighting, infiltration, and other management challenges. This compromises their organizational integrity. Interestingly, the indicators for groups emerging from unidimensional and multidimensional ruptures reach statistical significance in every model regardless of which variables are introduced. This is testament to the robustness of this relationship between the internal cohesion, preference alignment, and survivability of armed groups. It is also interesting that both indicators are significant in reference to *nonsplinter* organizations and not just to each other. This implies that the effects of splinter formation are not only limited to differences between splinters, subtly shaping their ability to survive, but rather that there is a larger, more general effect that is noticeable even among the entire population of militant organizations. And when it comes to specific types of disagreements, groups forming unidimensionally, over any issue, are much more durable than organizations forming multidimensionally, but those forming over strategy are the most robust of all.

Statistical Conclusions and Discussion

In this chapter I statistically evaluate rates of survival and radicalization among splinter organizations. I study a random sample of 300 armed groups operating between 1970 and 2012, and I find clear patterns linking how

Table 4.13 Summary of statistical findings

	Attack Frequency		Fatalities		Attack Lethality	
	Year 1	Total	Year 1	Total	Year 1	Total
Splinter Groups	.	▼	▲	.	.	.
Splinter—Strategy Involved	.	.	▲	.	▲	▲
Splinter—No Strategy	.	▼	.	▼	▼	.
Unidimensional Strategic	.	.	▲	.	▲	▲
Unidimensional Ideological	▲	.	▲	.	.	.
Unidimensional Leadership	.	.	▼	.	▼	.
Multidimensional	▼	.	.	.	.	.

Showing statistical estimates at $p < .1$

breakaway groups form to their short and long-term organizational trajectories. The results are summarized in Table 4.13.

Why do some splinter groups radicalize more than others? I find that radicalization most commonly follows internal disagreements over strategy. Whether unidimensional or in combination with other issues, these splits are linked with increased radicalization in half of all the analyses I conduct. More specifically, I find that breakaway groups motivated by strategy generate more fatalities and a higher per-attack lethality in their first year (+176% and +124%, respectively), and then sustain their elevated per-attack lethality (+44%) throughout their entire lifespan. No other group comes close to matching their consistent embrace of violence from their first days to their last.

Otherwise, some interesting and unexpected patterns emerge. For one, there is evidence that all splinter groups cause more fatalities in their first year of activity compared to similar nonsplinter groups. As noted in previous chapters, breakaway groups face some distinct challenges upon forming: not only must they establish their own reputations, their own sources of funding, and their own recruitment networks, but they must do so in the shadow of their parents. This could lead to an especially acute form of outbidding that motivates even more severe acts of violence. This is also one of the first analyses to explore variation in operational patterns as armed groups mature. While existing research commonly takes age into account, it is usually as independent variable and not to stratify samples for the sake of uncovering changes over time. In the future, more research should be conducted on how operational patterns evolve across the life-cycles of armed groups.

In addition, I find that splinter groups emerging from ideological disputes show a proclivity for suicide bombing. In cases where ideological splits occur, it may be that ideology is particularly central to the group's identity. It could also be that ideological hardliners are most likely to break away, and their ideological fervor could explain their predilection for suicide bombings. Since my coding scheme does not differentiate between ideological ruptures concerning politics, religion, or economic doctrine, it would useful to explore whether religious fractures have distinct effects.

Of course, some splinter groups do exhibit lower rates of radicalization. This is primarily the case for splinters motivated by leadership disputes and, to a lesser extent, those forming multidimensionally. Since there is no existing research that directly bears on this question, it is difficult to say for certain why this pattern emerges. But my theory, and the case studies, do offer some explanations. First, disagreements over leadership and control may produce armed groups that prone to fighting with their parent organization, while multidimensional splits produce groups prone to fighting among its own members. These inter- and intra-group attacks might not be included in the data sets I use for this study—data sets that primarily focus on violence against civilian populations—leading to the perceived lower rates of activity. Second, it could be that groups emerging from leadership and multidimensional fractures exhibit lower rates of internal cohesion. While this is most anticipated for the latter, groups forming around a single figure may also have lower organizational capital to draw upon. This could undermine their ability to sustain operational tempo, to plan complex attacks, and to execute missions that require coordination and cooperation. In effect, the preference divergence and internal problems these groups experience is likely working against their attack capacity.

Next, why do some splinter groups survive while others quickly fail? These results are summarized in Figure 4.9. In line with my theory, the statistical results reveal that splinter groups are more nuanced than current research suggests. Groups forming unidimensionally around a single issue, and with more aligned internal preferences as a result, are the most durable. The homogeneity of their membership base decreases rates of internal feuding and defection, leading to more effective control and greater organizational integrity. On the other hand, splinters forming multidimensionally experience greater preference divergence, more feuds, and lower rates of cohesion and unity. I find these organizations to be the least durable of all.

More specifically, across every statistical model presented in this chapter, the data reveal that splinter organizations from multidimensional ruptures were consistently and significantly more likely to fail at any given point in time. This finding cannot be attributed to unique operating conditions or differences in group ideology since I control for a range of other relevant factors that might simultaneously influence their ability to survive. I also find that unidimensional fractures do not generate monocausal effects. When disaggregated, slightly different patterns of survival emerge. This is depicted in Figure 4.9, which plots the relative, expected rates of survival for different organizations. Overlap indicates that estimates are not statistically different. Thus, strategic splinter groups are more durable than multidimensional and nonsplinter groups, but they overlap with ideological and leadership splinters.[w] Splinters that form over strategic disagreements are therefore some of the most durable organizations of all. However, any type of unidimensional fracture produces groups that are more durable than those emerging from multidimensional fractures, as my theory anticipates.

Why might strategic schisms produce groups that are particularly durable? It could be that these fractures create organizations that are the most aligned with respect to their actual, day-to-day behavior.[x] This is especially important because disagreements over strategy can be particularly divisive since

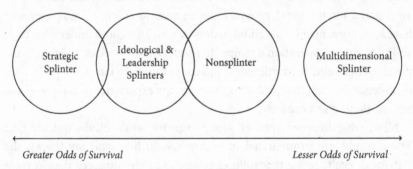

Greater Odds of Survival   Lesser Odds of Survival

Fig. 4.9 The relative survival rate of different groups

[w] In other words, post-tests reveal that their coefficient estimates are not statistically different at the $p<.1$ level.

[x] It could be that these members also exhibit greater commitment, with their dedication to violence proving how committed they are to their objectives. However, existing research would suggest that ideological, and not tactical, commitment is important to group survival.

operatives see them as matters of life and death. In contrast, when militants form around shared ideological or personal preferences, they do not necessarily hold similar views that guide their behavior. Though groups might share dissatisfaction with their parent organization's commitment to religious or political doctrine, there might still be underlying disagreement over their strategic and tactical choices that generate tension and undermine cohesion. Consequently, the type of preference alignment that occurs when militants fracture over strategic differences may be most beneficial to their cohesion and, ultimately, their survival.

Together, there are three main conclusions from this chapters. First, rates of survival and failure vary across splinter groups according to the logic of their initial formation. Understanding why and how groups form is therefore crucial to understanding the likelihood they will survive and pose a threat to both states and civilians. While splinter groups emerging from multidimensional ruptures are less stable, those from unidimensional ruptures are quite durable. From a policy perspective, fracturing armed groups should therefore not be seen as an unequivocal victory. On the contrary, doing so can create new organizations that are even more durable than the groups they depart. Instead of blindly fomenting internal feuds, policymakers should carefully consider the long-term ramifications of their actions and avoid fomenting ruptures that rally individuals under a common banner.

Second, rates of radicalization likewise vary across splinter groups and in line with their motivations for breaking away. When strategic disputes are to blame, splinter groups are most likely to ramp up violence. They generally form over a shared desire to renew or expand the use of violence, and evidence shows that they follow through on these goals. As before, this has important implications for policymakers. It again implies that blindly fragmenting one's opponents is unwise. Fomenting strategic disagreements is particularly dangerous since it can produce organizations capable of launching some of the deadliest attacks. However, this is not to say that fragmentation can never be a useful counterinsurgency tactic. Actions that lead groups to split up multidimensionally or over leadership disagreements appear most likely to produce weaker *and* shorter-lived breakaway groups. By fragmenting militants in this way, the resulting splinter organizations will be relatively weak, while the parent organization will lose part of its membership base.

Finally, from a broader perspective, these finding confirm that the conditions underlying a group's formation profoundly impact their long-term trajectory. Understanding how and why armed groups form can help law-makers and academics alike to more fully understand and predict how they will behave. In particular, this research has profound implications for groups that emerge during periods of negotiation or ceasefire. Many of these, like the Continuity IRA and the Real IRA, broke away over strategic differences, attracted disaffected hardliners, and consequently became more dangerous than their predecessors. When such possibilities are likely—for instance, during negotiations with rebels—policymakers should consider how they can prevent these ruptures and particularly hardline splinters that could undermine their progress.

5

Al-Qaeda and the Islamic State

Introduction

One of the most prominent and debated organizational ruptures in recent history took place in 2014 among the ranks of Al-Qaeda (AQ), the deadly terrorist organization responsible for the September 11th attacks on the United States. While this attack established Al-Qaeda as one of the world's most notorious armed organizations, its split produced a rival—the Islamic State (IS)—that, in many ways, superseded its operational capacity and raw brutality.

In this chapter I apply my theory to understand IS, tracing its evolution from the fevered imagination of a Jordanian misfit to a state-like bureaucracy overseeing nearly 34,000 square miles of territory.[1] Methodologically, this is a deviant case: it is a highly unusual instance of organizational splintering and it is subsequently challenging to directly apply my theory, to test its mechanisms, and to explain the full course of events. This is because IS did not form as a subgroup within AQ, as splinters typically do, but it was a distinct entity that merged with and then later broke away from it. In effect, AQ essentially imported a faction and began at Stage Two of my theoretical framework. My goal with this case study is therefore not to conduct a rigorous test of my theory as I did with splinter groups in Northern Ireland, but to understand how factional politics and intragroup dynamics shaped the course of events and reveal lessons for the future.

Evaluating IS is also important because its independence and quick ascendance caught the international community off guard. As Lahoud and Collins explain, the emergence and growth of IS was facilitated by the United States' Authorization for the Use of Military Force (AUMF), passed in the wake of the September 11th terrorist attacks.[2] The AUMF provided US armed forces broad leeway to hunt down those responsible for the attack, but it "served as an incentive to lump regional jihadi groups under the AQ umbrella" and to view all armed groups as a single, monolithic threat. But my theory offers an additional explanation: analysts and researchers often employ group-

Divided Not Conquered: How Rebels Fracture and Splinters Behave. Evan Perkoski, Oxford University Press.
© Oxford University Press 2022. DOI: 10.1093/oso/9780197627075.003.0005

based frameworks to understand armed organizations. They are viewed as unitary actors while ignoring much of what goes on inside.[3] This explains why so many overlooked the internal politics and intraorganizational feuds that portended the Islamic State's independence and that should have given forewarning of what was to come.

Applying my theory to the Islamic State offers other benefits as well. It demonstrates how US actions in the Middle East impacted the evolution and inner workings of armed groups in unpredictable ways. Existing approaches to counterterrorism and counterinsurgency are predicated on how policies might contribute to an organization's defeat, but they ignore what other outcomes may be possible: outcomes ranging from the emergence of hard-line splinter groups to new leaders taking control to alliances between former enemies attempting to survive. The history of the Islamic State offers important lessons for understanding externalities like these.

Abu Musab Al-Zarqawi and the Formation of Jama'at Tawhid wal-Jihad

While a single chapter is not enough to cover the full complexity of IS, my goal here is to provide enough detail for an accurate, informed analysis of how it emerged, why it joined Al-Qaeda, why the two organizations split, and how the Islamic State has since developed on its own.

Central to the story of the Islamic State is a Jordanian named Abu Musab Al-Zarqawi. Often described as a charismatic, natural leader with a long criminal past, he had no real military credentials and no significant interest in religion in his early adulthood. Weiss and Hassan describe him as an "unpromising student who wrote Arabic at a semi-literate level, dropped out of school in 1984, the same year his father died, and resorted immediately to a life of crime."[4] This led to his first of two stints in jail, this time for drug possession and sexual assault.

Zarqawi's life transformed in the late 1980s after his mother pressured him to attend services at a local mosque in Amman. He became engrossed and "plunged into Islam with all the passion he had once reserved for his criminal pursuits."[5] It is no surprise that when leaders at the mosque asked for volunteers to fight in Afghanistan alongside the Mujahideen, he jumped at the chance to act upon his newfound devotion. He arrived in Afghanistan at the tail end of the Soviet's withdrawal, with the Communist government still standing but no longer backed by Soviet forces. Here, he gained valuable military training and combat experience while fighting to retake towns and villages throughout the countryside. This experience had a profound effect. "[Zarqawi] had ... drunk from the heady cocktail of battlefield camaraderie and the rebels' own improbable success."[6] With his experience on the battlefield only reinforcing his recently developed religious fervor, Zarqawi's experience in Afghanistan left him with much ambition for a future that would combine religion and war.

Zarqawi returned to Amman after three years and he quickly became dissatisfied with—in his eyes—Jordan's overly liberal social and political climate. Frustrated, he joined up with an amateur group of extremists who called themselves Loyalty to the Imam.[7] They were plotting an attack that was quickly uncovered by the intelligence services. Each member was arrested and Zarqawi ended up in prison again, but this time for nearly five years. As with his combat experience in Afghanistan, prison was another formative experience. "Prison was his university," according to one US counterterrorist official.[8] And from this university, where he amassed a loyal following, Zarqawi emerged as a more potent, determined, and effective leader. Upon his release, he left Jordan for good, going first to Pakistan and then on to Afghanistan once again.

Zarqawi's second move to Afghanistan is perhaps most important because it led to his first meeting with Osama Bin Laden in 1999, the emir of Al-Qaeda and leader of the international jihadist movement. Upon arriving in Afghanistan, Zarqawi's goal was to establish his own jihadist training camp where he could prepare fighters for operations throughout the Middle East. He found himself with little in the way of funds or connections, but Bin Laden soon intervened on his behalf. Bin Laden leveraged his connections with the Taliban so Zarqawi could set up his own base of operations in the southern province of Herat and provided some initial training to Zarqawi and his band of fighters. He also provided seed money directly from Al-Qaeda coffers to help Zarqawi get started.[9]

Once it was up and running, Zarqawi's camp operated largely independently of Al-Qaeda's more substantial operations in the country. Initially, there was no indication that the two organizations would ever collaborate more significantly. As I discuss later in this chapter, Bin Laden held strong reservations about Zarqawi, and for his part, Zarqawi resisted Bin Laden's request to swear allegiance to Al-Qaeda.[10] Nonetheless, "The lopsided deal favored Al-Zarqawi; he received critical assistance but maintained his independence, all while embracing radical jihadis that complicated al-Qa'ida's political position in Afghanistan."[11] This new group of jihadis, many of whom would have been turned away by Al-Qaeda for criminal tendencies or doctrinal differences, soon started to call them themselves Jund-Al-Sham. They later became known Jama'at Tawhid wal-Jihad (JTJ) and Abu Musab Al-Zarqawi was their leader.[a]

Soon, the politics of the greater Middle East, and Afghanistan in particular, would be fundamentally altered with the spectacular terrorist attacks of September 11, 2001. Al-Qaeda organized and executed these attacks from its base in Afghanistan, with airplane hijackers killing nearly 3,000 American citizens and destroying the iconic Twin Towers at the World Trade Center in Manhattan. In response, the US invaded Afghanistan in 2001 and then Iraq in 2003. In Afghanistan, US forces routed the Taliban government that was providing safe haven to Al-Qaeda. Without this support, Al-Qaeda would have been unable to operate the training camps that were critical to the success of the 9/11 plot. The initial phases of the US invasion easily succeeded, the Taliban was ousted from the capital city of Kabul, and the insurgents they sheltered quickly fled to other countries. Many of them, including elements of Al-Qaeda and Zarqawi's JTJ, absconded to Iraq around 2003. Others, including Al-Qaeda's senior leadership, went instead to Pakistan and narrowly avoided US special forces, mostly famously in the Battle of Tora Bora in December 2001.[12]

---

[a] An obvious question is why Bin Laden agreed to help Zarqawi in such a lopsided deal. First, some scholars suggest that the alliance, even if imperfect, prevented Zarqawi from aligning with (and thereby strengthening) other jihadist networks in Afghanistan, particularly that of Abu Musab Al-Suri (Fishman 2016a, p. 31). Al-Suri was a rival jihadist leader who had defected from Al-Qaeda in 1997 over strategic differences (specifically over AQ's decision to target Western powers). Once independent, he began operating more closely with the Taliban (Hamming 2019b). Second, and perhaps more significantly, Zarqawi provided Al-Qaeda with a potential foothold into the Levant— the area roughly comprising Syria, Jordan, Israel, Palestine, and Lebanon. At that point in time, Al-Qaeda operations were largely confined to Afghanistan and the group had little infrastructure, personnel, or connections in other areas. Zarqawi, however, provided a natural segue into these new domains as he grew up in Jordan and was of Palestinian descent. Of course, this connection was complicated by the fact that Zarqawi was not necessarily loyal to Al-Qaeda, and his JTJ operated outside of Al-Qaeda's chain of command. Third, some scholars argue that Bin Laden's decision was at least partially swayed by his trusted security chief, Saif Al-Adel, who was especially keen on Zarqawi (Weiss and Hassan 2016, p. 12).

Less than two years after invading Afghanistan, US forces invaded Iraq over misguided fears that Saddam Hussein was pursuing weapons of mass destruction. As with Afghanistan, the initial invasion was a success. But problems soon emerged. There were few plans for how to govern and rebuild the country once Saddam Hussein's Ba'athist regime was toppled, and the US occupying forces' ad hoc plans only inflamed ethnic and religious tensions in Iraqi society—tensions that the brutal dictatorship of Saddam Hussein had suppressed. What came after initial military successes, in both Afghanistan and Iraq, was much more challenging: two protracted insurgencies that the US military was neither prepared nor trained for.

Zarqawi, still in Afghanistan, saw the Iraqi insurgency as an opportunity. In 2003, JTJ joined the fray to fight alongside Iraqi Sunni insurgent groups like the 1920 Revolution Brigade and Ansar-Al-Islam. But unlike his Iraqi counterparts who were mostly interested in resisting the US invasion and establishing a new government in post-Saddam Iraq, Zarqawi viewed this conflict in broader religious terms, which created tension between him and his local allies. As he noted in 2006:

> We will fight in the cause of God until His shariah prevails. The first step is to expel the enemy and establish the state of Islam. We would then go forth to reconquer the Muslim lands and restore them to the Muslim nation.... I swear by God that even if the Americans had not invaded our lands together with the Jews, the Muslims would still be required not to refrain from jihad but go forth and seek the enemy until only God Almighty's shariah prevailed everywhere in the world.... Our political project is to expel this marauding enemy. This is the first step. Afterwards our goal is to establish God's shariah all over the globe.... We will not be revealing a secret when we say that we seek to establish Islamic justice in the entire world and crush the injustice of disbelief.[13]

In contrast to other Iraqi organizations whose efforts were principally directed against the American occupiers, and to a lesser extent their Shiite rivals, Zarqawi's motivations extended well beyond the immediate political environment. He viewed the US invasion as part of a prophecy that would lead to a worldwide Islamic caliphate. In effect, "The Jihadis [like JTJ] were not simply trying to roll back the invasion; they envisioned a radically different Iraq and not a return to the status quo ante."[14] Zarqawi's vision provided an important preview of the Islamic State's eventual ambitions, and it showed how the group's perspective would differ from other organizations in Iraq, including groups that it would eventually cooperate with. One Congressional Research Report indicated that "[Zarqawi's organization] is the only insurgent group in Iraq...with stated ambitions to make the country a base for attacks outside Iraq."[15] As a result, forces were set in motion that would inevitably lead to conflict among the armed groups fighting in the Iraqi insurgency.

Zarqawi's first few years in Iraq were marked by his penchant for recruiting foreign fighters. On the one hand, many of Iraq's native Sunni insurgent groups were comprised of local fighters: specifically, those who grew up in Iraq and were motivated to fight for Iraq's freedom. JTJ, on the other hand, preferred foreigners. By some estimates, nearly half of their fighters came from abroad. This matters for several reasons. Zelin, for instance, notes that JTJ's foreign fighters were especially attracted to how Zarqawi forced his interpretation of Sharia law "down people's throat."[16] This type of internal support undoubtedly contributed to Zarqawi's group becoming even more ruthless. As a result, JTJ found itself increasingly isolated from local organizations that were fighting for different reasons and had starkly different interpretations of their shared religious faith. Zarqawi's penchant for relying on foreigners was also a strong indication of what was to come, as future incarnations of his organization would also rely on outsiders to a significant degree.[b] Likewise, Zarqawi's lack of concern for resolving obvious disagreements with his peers is another enduring characteristic of his leadership.

If this period sheds light on the eventual goals and priorities of the Islamic State, which did not yet exist at this point in time, it also helps explain the organization's eventual operational profile. This is because JTJ's tactical profile bears a striking resemblance to the tactical profile of the Islamic State, and tracking the group's evolution could have given outside observers important clues about what to expect when the group exited Al-Qaeda. Specifically, JTJ engaged in suicide attacks, beheadings, indiscriminate violence against Shiites and their holy sites, and online propaganda to publicize its attacks and promote its goals. Rather than acting in accordance with its peers, JTJ had a distinctive operational profile.[17] As one Middle East Policy Council report notes:

[b] This trend indicates a tendency toward path-dependency among armed groups that analysts often fail to consider. That is, short-term decisions can affect the long-term trajectory of the organization.

JTJ differed considerably from the other Iraqi insurgent groups. Rather than using only guerrilla tactics—ambushes, raids and hit-and-run attacks against the U.S. forces—it relied heavily on suicide bombers. It targeted a wide variety of groups: the Iraqi security forces, Iraqi Shiite and Kurdish political and religious figures, Shiite Muslim civilians, foreign civilian contractors, and UN and humanitarian workers.[18]

JTJ's atypical operational profile is further reflected in attack statistics from the Global Terrorism Database. Comparing JTJ to its local contemporary and collaborator, Ansar Al-Islam,[c] JTJ conducted 15 times more suicide attacks, nearly 12 times more non-suicide attacks, and caused 15 times more fatalities and injuries between 2003 and 2006. The two organizations were operating at fundamentally different levels of warfare and, as discussed earlier, had radically different goals as well. As was the case with ideological distinctions, these operational distinctions between JTJ and other insurgent organizations should have been red flags to the international community. Instead of ignoring these differences and treating the Sunni insurgency as a monolithic threat, the US and its military partners in Iraq would have been wise to treat the insurgency as a complex and interdependent organizational challenge that would necessitate a sophisticated approach.

Zarqawi's JTJ continued along its brutal path, maintaining a significant and distinctive presence within Iraq's domestic insurgency. But an impending alliance would soon divert the group's trajectory and accelerate its evolution into an even deadlier force.

Becoming Al-Qaeda in Iraq

When Zarqawi and JTJ relocated from Afghanistan to Iraq in 2003, they maintained their links with Al-Qaeda but also maintained their independence. In August 2004, that link grew even stronger when Zarqawi agreed that his organization would become the newest Al-Qaeda franchise.[19] He changed the group's name to Tanzim Qaidat Al-Jihad fi Bilad Al-Rafidayn, which translates to Al-Qaeda in the Land of the Two Rivers. (i.e. Iraq, where

---

[c] GTD does not specifically refer to JTJ but instead to Al-Qaeda in Iraq. I discuss this name change in the coming pages.

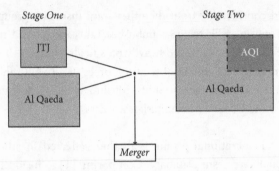

Fig. 5.1 JTJ becomes AQI

the Tigris and Euphrates flow). Over time, they became known simply as Al-Qaeda in Iraq (AQI).[20]

Figure 5.1 displays this merge in terms of my theoretical framework. As one can already see, this it is not a typical case: the organizational ruptures I conceptualize occur when a faction forms within existing armed groups and then break away. But here, the faction was essentially imported to the organization. While this is unusual, in many ways it did operate like a faction within AQ. As I discuss later, AQI had a distinct ethos, outlook, and leadership cadre, and after a few years, it also sought a fundamental transformation of the status quo. In this way, the franchising strategy that AQ employed here (and elsewhere) seemingly infused intraorganizational tension from the start, making the management of that tension a perpetual challenge.

To understand how this merger transpired after Zarqawi had refused to pledge loyalty to Bin Laden only a few years earlier, we must take stock of Zarqawi and Bin Laden's competing perspectives. What did each stand to gain and potentially lose? This is important because armed groups are commonly motivated by self-interest, causing them to compete over finite pools of funding, recruits, and weapons.[21] But in the case of JTJ and Al-Qaeda, these selfish desires were somehow tamed and the two managed to cooperate.

From Bin Laden's perspective, evidence suggests that he viewed his partnership with Zarwqawi through an overwhelmingly political lens. First, he wanted to bring Al-Qaeda into Iraq, and merging with Zarqawi was a low-cost means of establishing a permanent base in the country. As Zelin notes, "Bin Laden himself wanted to 'own' the Iraq jihad as well as remain relevant while hiding from the United States."[22] The alliance helped accomplish this goal: it placed the Al-Qaeda brand front and center within the Iraqi insurgency, at the heart of the Middle East's preeminent conflict with its superpower nemesis. Yet, they would not need to travel to Iraq, send their own forces, establish their own operations, or risk their own advisers. They could instead accomplish all of their goals by onboarding an existing insurgent operation. It was thus low-risk but potentially high-reward.[d] Second, Al-Qaeda was on the run from US and coalition forces, so the partnership with AQI could help renew its tattered image. Although Al-Qaeda was evicted from Afghanistan and largely hiding out in Pakistan, operations attributed to this new group, known as *Al-Qaeda in Iraq*, would bring prestige and renewed interest to AQ as a whole. Third, Bin Laden saw his partnership with JTJ as a means to unify the Iraqi Jihadist movement that was already showing major signs of division by 2004. Infighting, turf wars, and a lack of hierarchy and control plagued these insurgent organizations, and this only benefited adversaries like the United States. Data from the Global Terrorism Database, for instance, reveals that 20 different armed groups conducted attacks within Iraq in 2004 alone, which is likely a low estimate. Given these divisions, Bin Laden saw Zarqawi as a potential solution: "If Jihadis there were going to unify, Zarqawi would need to lead the movement."[23] And if Al-Qaeda were able to bring Zarqawi under its organizational umbrella, it could effectively claim the top spot for itself.

Yet, this partnership did pose some risk to Bin Laden and Al-Qaeda, perhaps most significantly from a brand perspective. If AQI were to fail, it would reflect poorly on Al-Qaeda as a whole. And if AQ were still on the run when this happened, with little capacity to conduct attacks and regain public attention, its prestige would surely diminish. Bin Laden may have believed that the Al-Qaeda brand could survive even an Iraqi defeat. In addition, there were security fears looming in the background: infiltration into one group could weaken both. Whether an intercepted communication or a defecting agent, some leak could imperil both organizations. Not surprisingly, this potential problem commonly deters substantial cooperation between armed groups.[24] With its experience in managing transnational relationships, however, Al-Qaeda may have felt capable of managing the

---

[d] Interestingly, the cooperation was low risk in terms of known-unknowns, but perhaps high risk in terms of unknown-unknowns. That is, no one could have suspected how JTJ would evolve into the Islamic State, whose success—unimaginable in 2004—has since threatened the Al-Qaeda brand. See, for instance: Charles Lister (Jan. 2016). *Jihadi Rivalry: The Islamic State Challenges Al-Qaida.* The Brookings Institution.

risk. And underlying these risks were fundamental differences between Zarqawi and Bin Laden—differences that unsettled Bin Laden from the start. Nonetheless, these concerns were not enough to deter Bin Laden, who ultimately gave his blessing and supported this new relationship.

From Zarqawi's perspective, there are some obvious downsides to tying his organization to Al-Qaeda and pledging allegiance to Bin Laden. First, his reputation was now linked to Al-Qaeda's successes and failures. AQ had been on the run, so it was not out of the question that it could suffer a mortal blow and leave an indelible stain on his legacy. While there was also a chance that AQI could be linked to AQ's successes, in 2004 these were relatively few and far between. Second, the merger would undoubtedly reduce Zarqawi's operational autonomy and infringe upon his personal independence. Since he would now be required to heed the advice of foreign commanders after pledging loyalty to Bin Laden, he could face new constraints in making tactical and strategic decisions. For a man with a known independent streak, this was no small matter. Third, if Zarqawi became dependent upon AQ's resources, he could easily end up being coerced. Disregarding Bin Laden's advice, for instance, could result in material restrictions such as stopping the flow of goods, fighters, or money to Zarqawi and his men. Finally, Zarqawi and Bin Laden were both aware of the security risks associated with managing such an alliance.

The merger ultimately went through, so we know that Zarqawi determined the rewards outweighed the costs. And the rewards were significant: "Among other immediate benefits for Zarqawi...was access to private donors and recruitment, logistics, and facilitation networks."[25] At the time of this merger, Al-Qaeda was perhaps the wealthiest jihadist organization in history, only to be surpassed by the Islamic State in the coming decade. Al-Qaeda earned several million dollars per year in the mid-2000s, although this was down from its yearly budget of nearly $30 million just prior to the September 11th attacks.[26] These funds could be funneled to AQI in support of Iraqi operations. Al-Qaeda was also well connected to smugglers and arms traders. Recall that Zarqawi was relatively new to Iraq when he arrived in 2003, so Al-Qaeda's connections would be useful. Additionally, a partnership with Bin Laden provided access to Al-Qaeda's explosives experts whose knowledge could be transmitted to Zarqawi's forces. All of this was enticing to Zarqawi who needed financial and logistical support to realize his aspirations—just as he had in Afghanistan a few years earlier.

While it is commonly glossed over by researchers and policymakers, it is important to carefully consider the structure of the relationship between Al-Qaeda and Al-Qaeda in Iraq. Was it a merger, an act of expansion or decentralization, or something else altogether? While this matters for the applicability of my theory, it also matters for accurately assessing the organizational dynamics of these groups.[e] In this sense, it is important to consider whether Zarqawi pledged bayat, or fealty, to Bin Laden, which would be a useful indicator of the relationship's strength.[27] If Zarqawi did pledge fealty and thereby subordinated his organizational autonomy to Bin Laden, it would indicate a high level of commitment between the groups—more of a merger than a simple collaboration. If not, then the AQ-AQI relationship would not be much different from other instances of sustained cooperation between armed groups that we commonly witness. This distinction also has important implications for how we understand the eventual disagreement between the two groups that caused them to part ways, and whether we conceptualize it as the breakdown of a cooperative alliance or the breakdown of a single group.

There is overwhelming evidence that Zarqawi *did* pledge bayat to Bin Laden. Zarqawi made this especially clear in a statement posted online in 2004:

It should bring great joy to the people of Islam, especially those on the front lines, and it was with good tidings of support during this blessed month that Tawhid wal-Jihad's leader, Abu Musab Al-Zarqawi (God protect him) and his followers announced their allegiance to the Sheikh Al-Mujahideen of our time, Abu Abdullah Osama Bin Laden, God protect him.[28]

This pledge of bayat effectively subordinated Zarqawi to Bin Laden,[29] and this is strong evidence that JTJ and AQ achieved an actual merger, not merely an agreement to collaborate.

The pledge of bayat brings up two additional considerations. First, while bayat is a *personal* and not *organizational* commitment, the pledge officially linked both organizations. That is, when the leader of one group pledges loyalty to another, it indelibly fuses them into a single hierarchy that effectively

---

[e] Regardless of the precise functional form, the general idea—that the drivers of factional dissent will shape the trajectory and composition of armed groups—should still help us understand the consequence of this rift.

cedes control to a single person. This meant that AQ and JTJ had essentially "joined ranks."[30] Even Zarqawi's wife hinted at this arrangement in her husband's eulogy, describing him as "nothing more than a soldier in one of the ranks of [Bin Laden]'s armies."[31]

Second, the practical relationship between AQ and AQI did not mirror what the pledge might otherwise suggest. In fact, there is little evidence that AQ and AQI operated like a single group. As Fishman notes, Zarqawi "[delineated] a distinct jihadi ideology and organizational ethos" that, in effect, maintained a stark boundary between the two entities.[32] This separation is also reflected in how the two groups did not share members seamlessly. While some fighters came to Iraq to assist AQI early on, there is not much evidence of personnel moving in the other direction. In addition, tactical decisions were almost entirely left to Zarqawi and local commanders, and even their strategic decisions (and anything related to activities within Iraq) were largely independent from Al-Qaeda as well.[33] As I discuss, one of the main disagreements between AQI and AQ leaders was over strategy, and AQ leaders had little success compelling Zarqawi to conform to their wishes.

Taken together, AQ and AQI occupy a peculiar space that does not fit neatly into existing typologies. What makes this arrangement especially difficult to categorize is that the relationship seems stronger on paper than it was in practice. This makes it challenging to identify analogies and to apply my theory. But ignoring these complexities in favor of parsimony or simplicity not only obscures the nuance that makes this case unique, but risks drawing incorrect inferences as well.[f]

Becoming the Islamic State

The net effect of becoming Al-Qaeda in Iraq was a temporary boon for Zarqawi and the former JTJ. With Al-Qaeda's help, AQI managed to control the flow of resources and fighters into Iraq between 2004 and 2006. This made them a major force in the insurgency, regardless of—or perhaps in addition to—its distinctive ideology and continued pattern of wanton violence.[34]

---

[f] I take this into account when considering how the groups separated in the following pages.

AQI's fortunes began to change around 2006. In a startling turnaround, the group was nearly routed by a coalition of US, Iraqi, and local Sunni forces.[35] A few factors were responsible for this unprecedented shift. First, the United States adopted a more effective counterinsurgency strategy. The Bush Administration ordered "a new way forward," moving from an enemy-centric to population-centric method of defeating insurgents, colloquially known as the Surge.[36] This new strategy proved highly effective.[37] Second, local armed groups turned against AQI and other foreign jihadist organizations in the Sons of Iraq Program (SOI, and unofficially known as the "Sunni Awakening"). Around this time, many Iraqi citizens became fed up with the brutal attacks and sectarian divisiveness. This prompted some of AQI's partners to withdraw their cooperation, actively oppose Zarqawi, and work with the Iraqi and American governments to defeat him.[38] In total, around 91,000 fighters took part in the SOI program by 2007. Combined with the US military's Surge forces, Zarqawi's enemies gained a decisive advantage.[39]

As a result, AQI was forced to abandon much of its position in southern Iraq where the Sunni Awakening was strongest. It retreated north to regroup in the Ninewa governorate, where it had residual strength from its earliest Iraqi operations starting in 2003.[40] For an organization that Bin Laden hand-picked to become the newest Al-Qaeda chapter only two years earlier, this retreat was a shocking and unwelcome turn of events. Not only did this setback make all of AQ seem weak, but the indigenous SOI movement "fueled the perception that Al-Qaeda, which had pledged to defend Muslim populations against foreign occupying forces, was itself an alien occupier."[41]

AQI slowly retreated from the public eye. With its dwindling capacity, it hunkered down around Mosul and engaged in sporadic acts of violence.[42] AQI also used this time to develop its relationship with other insurgent groups. Many of these, especially those with Iraqi roots, despised Zarqawi and the organization he had built owing to its independent streak and gratuitous violence. Things began to change in 2006 when, partially at the behest of senior AQ leaders, Zarqawi established the Majlis Shura Al-Mujahidin fi Al-Iraq, or the Jihadis Advisory Council in Iraq (also called the Mujahideen Shura Council). This was a coalition of six Sunni insurgent organizations, including AQI, designed to promote coordination and cooperation. Of course, this new cooperative approach was not motivated solely by goodwill; with Zarqawi's weakened position, he saw his peers as useful allies against US forces.[43]

AQI suffered a major blow in 2006 when a US drone strike killed Zarqawi. He was replaced relatively smoothly by Abu Hamza Al-Muhajir.[g] A well-known figure among the group's upper echelon with a reputation as an expert bomb-maker, Al-Muhajir's ascension came as little surprise.[44] But he is especially important to the AQ-AQI relationship for two reasons. First, he cemented and clarified their bond. Where Zarqawi apparently pledged bayat in private (later reiterating his commitment in an online statement), Al-Muhajir did so publicly and proudly. In his first statement as group leader, he communicated to Bin Laden that "we are at your beck and call and at your disposal"—a stronger sentiment than any Zarqawi conveyed.[45] And retrospectively, Ayman Al-Zawahiri (the commander of Al-Qaeda after Bin Laden was killed in 2011) reaffirmed Al-Muhajir's pledge in a letter from 2014, noting that he considered AQI to be a part of AQ and not merely a close collaborator.[46] Al-Zawahiri's statement is further evidence that AQ and AQI were effectively one organization.

Second, and in a fundamental shift, Al-Muhajir was in command of AQI in October 2006 when the Islamic State of Iraq (ISI) was created. While he neither proclaimed nor led this new state on his own, he was instrumental in its inception. The idea actually came from Al-Qaeda's number two, Ayman Al-Zawahiri, who saw the proclamation of an "Islamic state" as a way to broaden AQI's control, hijack Iraqi politics, and bring disparate Sunni tribes under the Al-Qaeda umbrella. In some ways, this represented an extension of the logic underpinning the initial creation of the Mujahideen Shura Council described earlier, though even more selfishly motivated.

The first leader of ISI was Abu Umar Al-Baghdadi, a relatively obscure figure who was leading the Mujahideen Shura Council at the time.[h] Al-Baghdadi was so unknown that US forces believed "Al-Baghdadi" was just a nom de guerre.[47] Nevertheless, in 2006 it was Al-Baghdadi who declared that "[We have] reached the end of a stage of jihad and the start of a new one, in which we lay the first cornerstone of the Islamic Caliphate project and revive the glory of religion."[48] While it was an overly bold announcement for a group with such little capacity, ISI would get much stronger in three weeks when Al-Muhajir pledged his 12,000 AQI fighters, and another 10,000 in training, to Baghdadi's control. And while the initial idea for the Islamic State came from Al-Qaeda, it was neither consulted about who would lead

---

[g] Al-Muhajir also goes by the name of Abu Ayyub Al-Masri.
[h] Note that this is not Abu Bakr Al-Baghdadi who would take over ISI several years later.

this new organization, the timing of Al-Baghdadi's proclamation, nor about Al-Muhajir pledging his troops entirely to Al-Baghdadi's control. From this point on, both Al-Baghdadi and Al-Muhajir claimed that AQI no longer even existed and that it had been completely replaced by ISI.[49] Undoubtedly, AQ leaders outside of Iraq were watching with concern as their plans spiraled out of control.

With this transformation in 2006, it is again worth considering the relationship between the newly created ISI and AQ. Was ISI a new, independent entity, or was it a continuation of Zarqawi's legacy—and therefore a continuation of the bond with Al-Qaeda? On the one hand, ISI leaders like Al-Baghdadi claimed they were never loyal to Al-Qaeda as they had never pledged their allegiance to Bin Laden. AQ leaders refuted this claim and maintained that ISI leaders had indeed pledged loyalty, albeit in private, and therefore were continuing the connection that was forged by Zarqawi. To corroborate this claim, Zawahiri released numerous correspondences between himself and ISI leaders, with one even asking him directly: "[D]o we renew the pledge in public or in secret as before?"[50] Even if ISI leaders like Al-Baghdadi did not make their own pledges of bayat, AQ argued that Al-Muhajir's pledge less than a year before would carry over to this new group. Even more evidence to support the notion that ISI leaders pledged bayat came from captured members of Al-Qaeda who, when interrogated, confirmed it. As one American member of AQ described:

After Sheikh Abu Mus'ab's martyrdom in 2006, leadership of his group was transferred to Sheikh Abu Hamza al-Muhajir (may Allah have mercy on him), a former member of the Egyptian Jihad Group, who soon announced (without consulting al-Qaeda's central command) the dissolution of the group and the formation of what was known as the Islamic State of Iraq under the leadership of Sheikh Abu 'Umar Al-Baghdadi (may Allah have mercy on him), which in turn declared its allegiance to al-Qaeda's central command, except that this time the bay'at (pledge) was kept secret at the request of the brothers in Iraq.[51]

While evidence supports the pledge of loyalty, there is no way to adjudicate between these competing claims.[52]

From 2006 to about 2010, the story of ISI is largely a continuation of AQI's experience from 2003 to 2006. ISI was hammered by air strikes and with the increased number of US forces in Iraq, its leaders were frequently killed or arrested. ISI's extreme violence alienated locals and its clashes with other insurgent groups left it isolated. Describing US success in Iraq in June 2010, US Army General Raymond Odierno noted that "Over the last 90 days or so, we've either picked up or killed 34 out of the top 42 Al-Qaeda in Iraq leaders. They're clearly now attempting to reorganize themselves. They're struggling a little bit. They've broken—they've lost connection with [Al-Qaeda Senior Leadership] in Pakistan and Afghanistan."[53] By some accounts, the group was on the verge of defeat. As one Congressional Research Report notes, however, "[ISI] remains lethal and has the potential to revive in Iraq."[54] The group managed to survive by going underground, rethinking its security arrangements, and decentralizing its organizational structure. As Fishman notes,[55] Al-Muhajir and Al-Baghdadi are often criticized for their leadership skills, but under their guidance, ISI nonetheless survived the loss of 80 percent of its commanders. This was no easy task.

The Second Al-Baghdadi and Turmoil with Al-Qaeda

Despite its efforts, ISI was not able to protect those at the very top of its organizational hierarchy, and in April 2010, both Al-Muhajir and Abu Umar Al-Baghdadi were killed. This paved the way for perhaps the group's second most notorious leader to emerge: Abu Bakr Al-Baghdadi. Holding a PhD in Islamic Studies from the Islamic University of Baghdad, he has been described as the intellectual engine of the Islamic State. His ambitions fundamentally shaped the group's course in the years to come. The new Al-Baghdadi took over at an opportune moment, just as ISI's fortunes were starting to improve. This was due to four factors, the most significant of which were beyond the group's control.

First, the Iraqi government under Nouri Al-Maliki inflamed sectarian tensions through a combination of crucial mistakes and intentionally divisive actions.[56] Maliki has been described as a dictatorial leader and his predominantly Shiite government harassed and murdered Sunnis, excluded them from political and military positions, and halted funding to the successful Sons of Iraq program.[57] In effect, Maliki's policies pushed many Sunni fighters into the waiting arms of ISI—fighters who had recently been employed by the government and now found themselves with no source of income. By some estimates, 40 percent of ISI fighters in 2010 had previously been part of Sons of Iraq.[58] Maliki's blatantly sectarian policies also had the secondary effect of making Iraqi Sunnis much more likely to tolerate and even support ISI.

Second, another factor in ISI's resurgence was the withdrawal of US combat troops from Iraq beginning in 2011. As mentioned earlier, the Surge—where US troops flooded Iraq as part of a new population-centric counterinsurgency strategy—was critical to turning the tide against AQI in 2007. Now, these troops were being pulled out just as Sunni sympathies were growing and as ISI was rebounding. This decision would prove to be a recipe for disaster and one that ISI was quick to exploit.

Third, ISI benefitted from the improved leadership of Abu Bakr Al-Baghdadi, who was much more capable than his predecessors, Abu Umar Al-Baghdadi and Abu Hamza Al-Muhajir. Notably, Abu Umar Al-Baghdadi (referred to in the rest of this chapter simply as "Al-Baghdadi") oversaw an impressive reorganization of ISI: he institutionalized much of the group's responsibilities by creating specific departments for combat, training, and so on. He also embraced unemployed fighters, commanders, and intelligence officers who served under Saddam Hussein. One of these individuals was Hajji Bakr, a former military officer in the Iraqi army who became the military commander of ISI.[59] Al-Baghdadi recognized the talent and the skills these individuals could bring to his organization, embracing them in a way his predecessors did not. Researchers argue that this decision was especially critical to ISI's rebirth and future success.[60]

Fourth, and finally, ISI's prospects were buoyed by the onset of the Syrian civil war.[61] Beginning in 2011, the war in Syria presented an unprecedented opportunity for ISI to expand into an area with little government control but with a population of receptive Sunnis who had long been neglected by the Syrian government. ISI did not move into Syria immediately, but it—along with Al-Qaeda—helped to establish a new group in Syria called Jabhat Al-Nusra (JN). The establishment of JN was typical of Al-Qaeda: instead of moving into new territory unilaterally, Al-Qaeda either created a new group (as it did with JN) or took over an existing group (as it did with AQI).[62] Over time, ISI and JN grew closer owing to their geographic proximity across the Iraqi-Syrian border, cooperating on operations and supporting one another. Soon, the United States would name JN as a subsidiary of AQI—even though AQI was really ISI at this point.[63]

In 2013, ISI unilaterally declared a merger with JN while simultaneously changing its name from the Islamic State of Iraq to the Islamic State of Iraq and Syria (ISIS).[64] These two groups, according to ISIS, were now jointly under the leadership of Abu Bakr Al-Baghdadi and not Ayman Al-Zawahiri, who took over Al-Qaeda after Bin Laden's death in 2011. This infuriated Zawahiri who released a brief but historic statement in February 2014:

> Firstly: Qae'dat Al-Jihad [Al-Qaeda] declares that it has no links to the ISIS group. We were not informed about its creation, nor counseled. Nor were we satisfied with it rather we ordered it to stop. ISIS is not a branch of AQ and we have no organizational relationship with it. Nor is Al-Qaeda responsible for its actions and behaviors. The branches of AQ are those that have been announced by the Central Command, those are the ones we acknowledge.[65]

This was a clear, stark rebuke that left no room for interpretation. This effectively terminated the long, complex relationship between AQ and ISIS, and the two were now wholly independent. But what specifically caused this turn of events?

Breaking the Link between Al-Qaeda and the Islamic State

Conceptually, we can distinguish between factors that triggered the split between ISIS and AQ—the factors that are most strongly and perhaps causally linked to the breakdown—and the latent disagreements that long plagued the two.[66] Separating our these latent disagreements is important because some of them were evident from Bin Laden's first meeting with Zarqawi and had been effectively managed over the years.[67] It would be incorrect to say that these disagreements caused the split; while they elevated tensions between the two organizations, and likely contributed to the possibility that one event could rupture their unity, there were other triggers that temporally preceded and ultimately caused them to part ways.

To begin with, there were a host of latent disagreements perpetually plaguing the relationship between AQ and ISIS. One had to do with religious interpretation, and this was evident from the first days of their relationship. As noted earlier, Bin Laden and Zarqawi met in Afghanistan in 2000 not long after Zarqawi was released from a Jordanian prison (his second stint in Afghanistan). Once there, Zarqawi needed Bin Laden's funding and connections, but it was immediately clear that the two did

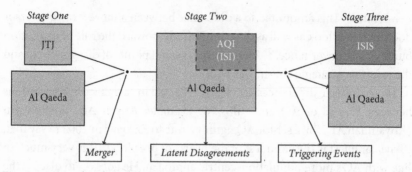

Fig. 5.2 ISIS splits from Al-Qaeda

not see eye to eye on matters of jihadist ideology. In fact, Bin Laden left their first meeting lamenting Zarqawi's overly "rigid views."[68] As Hamming confirms, the Islamic State's distinct ideology (and culture)"[infused] a critical prospect for internal tensions into the inter-group relationship from the very beginning."[69] Bin Laden's displeasure at Zarqawi's ideology was so strong that he nearly rebuked the initial request for assistance but was persuaded otherwise by Saif Al-Adl, a senior Egyptian Al-Qaeda leader. Al-Adl managed to convince Bin Laden to lend his support "despite all parties being well-aware of the ideological and theological differences between al-Qa'ida and the more hardline al-Zarqawi and the people around him."[70] This point underscores how researchers often assume homogeneity among groups under the same broad ideological umbrella, but their differences can actually be quite meaningful.[71]

In addition to matters of ideology, there were major differences in strategy. This revolved around fundamentally competing notions of how to achieve victory. While the two groups held largely similar objectives—the establishment of an Islamic caliphate in the Middle East—they disagreed over how to achieve it. On the one hand, ISIS pursued a fast-paced strategy that largely ignored popular support. It sought to impose a state rather than to build one, and violence was the primary tool. ISIS also wielded violence to purge its territory of non-believers (non-Muslims, Shiites, and non-devout Sunnis), which was—in its view—a critical first task.[72] AQ, on the other hand, waged a slower, more population-centric battle to win local support and expand outwards. Instead of concerning itself with local non-believers and less devout Sunnis, AQ was more interested in taking on the region's wayward governments. This, it hoped, would help attract supporters. As

Zelin notes, "This amounted to a difference between a more strategic versus doctrinaire outlook as well as differing attitudes toward the role of institution building and governance."[73] These were no small points of disagreement and the resulting strategic chasm was wide.

These strategic differences are clearly reflected in several communications between leaders of AQ and ISIS. For example, Atiyah Abd Al-Rahman (Atiyatullah Al-Libi; a senior AQ figure) wrote to Zarqawi in 2005 to say that, "True conquest is the conquest of the hearts of people." This is very much in line with AQ's more population-centric approach. He goes on to discuss the merits of:

> ... embracing the people and bringing them together and winning them over and placating them and so forth, for this, my brother, is a great way towards victory and triumph that is not lesser than military operations, but rather in truth is the foundation while military operations must be a servant that is complementary to it. Therefore, when you embrace the people and enjoin them through your morals, kind words, your conduct and upbringing, you will have gained a greater means of victory over your enemy, with God's permission.[74]

Notably, Al-Rahman takes care to describe AQ's strategy not as a replacement for military operations, which were critical to ISIS's identity and recruitment efforts, but as a supplement to them. In other words, he was not arguing that ISIS fundamentally abandon the use of violence, but rather utilize it in a more calculated, controlled manner that would not repulse and potentially isolate locals. Later in the same letter to Zarqawi, however, Al-Rahman gets more blunt with his language, writing "Let us not merely be people of killing, slaughter, blood, cursing, insult, and harshness; but rather, people of this, who are unopposed to mercy and gentleness." He adds that "Policy must be dominant over militarism.... That is to say, that military action is a servant to policy."[75] While ISIS leaders never fully internalized this message, by the late 2000s they did take a page from AQ's playbook to build government institutions and provide some social services. Nonetheless, they continued to perpetrate wanton violence and launch indiscriminate attacks.

AQ and ISIS obviously advocated very different strategies, which unsurprisingly led to very different actions on the ground. These actions inflamed tensions even more, with three sets of actions causing especially deep rifts between the two groups.

First, AQ was extremely displeased with ISIS's relentless attacks on Shiite Muslims and some of their holiest sites in the Middle East.[76] For example, IS reportedly killed 1,700 unarmed Shiite cadets from the Iraqi Army and 670 Shiite prisoners from Mosul over the course of 2014. The group also destroyed innumerable Shiite mosques, shrines, and cultural symbols as well, one of which was Samarra's famed Al-Askari Mosque with its golden dome.[i] As to why, Zarqawi and later Al-Baghdadi believed IS would benefit from a full-blown sectarian war. Such a war, they imagined, would drive those Sunnis who had refused to take sides into their ranks. This would become a core ISI strategy, and it was especially clear from a letter written by Zarqawi to leaders of AQ in 2004 that was intercepted by US forces. As he writes:

> [The Shiite] in our opinion are the key to change. I mean that targeting and hitting them in [their] religious, political, and military depth will provoke them to show the Sunnis their rabies and bare the teeth of the hidden rancor working in their breasts. If we succeed in dragging them into the arena of sectarian war, it will become possible to awaken the inattentive Sunnis as they feel imminent danger and annihilating death at the hands of these Sabeans.[j] Despite their weakness and fragmentation, the Sunnis are the sharpest blades, the most determined, and the most loyal when they meet those Batinis (Shiite), who are a people of treachery and cowardice.[77]

Unfortunately for AQ, Zarqawi's strategy paid off. As Ingram, Whiteside, and Winter argue, "[Zarqawi's] sectarian focus and prolific terror campaign—not to mention the Shi'a death squads that hit back—pushed thousands of fence-sitting Iraqis into his ranks."[78] And while many Sunni groups benefited from this influx of new fighters, few had the reputation of being as anti-Shiite as ISIS, which made it a popular destination for Sunni fighters.

AQ leaders repeatedly tried to rein in ISIS's attacks on Iraqi Shiites. Soon after AQI was officially announced, Bin Laden wrote in a letter that Zarqawi had "clear guidelines to focus his fighting against the occupying invaders, headed by the Americans, and to spare all those who maintained neutrality."[79] Then, once ISIS attacks on Shiites began in earnest, another

---

[i] Thankfully, the mosque reopened in 2009.

[j] The Sabeans were a group of people living in the southern Arabia peninsula around the time that the Quran was written, though more recent Islamic scholarship has traced their roots to people living presently in southern Iraq. In this context, it refers generally to non-Muslims, particularly Shiites, in Iraq.

letter was sent by a different Al-Qaeda commander (Atiyah Al-Rahman) who again argued with Zarqawi but from a purely strategic perspective. In his letter to Zarqawi, Al-Rahman wrote:

> And if the attacks on Shiite leaders were necessary to put a stop to their plans, then why were there attacks on ordinary Shiite? Won't this lead to reinforcing false ideas in their minds, even as it is incumbent on us to preach the call of Islam to them and explain and communicate to guide them to the truth? And can the mujahedeen kill all of the Shiite in Iraq? Has any Islamic state in history ever tried that?

Taken together, the decision to attack Shiites reflects the major strategic divide between the groups, with IS focused on purging the local community at all costs and AQ focused on overthrowing apostate regimes and not on disputes with fellow Muslims.[k]

Second, AQ took issue not only with ISIS attacks on Shiites, but with attacks on Sunni civilians as well. While AQ was unhappy about the level of brutality directed towards Shiites, Sunnis were different—they were AQ and ISIS's religious kin. Accordingly, AQ was careful not to attack or to kill Sunnis unnecessarily so as not to alienate its core constituency.

ISIS, however, was dedicated to purging the local Islamic community of those who they deemed to be non-believers. While AQ and many jihadist groups considered Shiite to be takfiri (that is, non-believers or non-Muslims), ISIS was one of the few to view Sunnis it deemed insufficiently religious in the same way, justifying their brutal attacks. As noted earlier, AQ was loath to support such attacks, urging ISIS to embrace Sunnis even if imperfect. As Zawahiri wrote to Zarqawi in 2005:

> If we look at the two short-term goals, which are removing the Americans and establishing an Islamic [emirate] in Iraq, or a caliphate if possible, then, we will see that the strongest weapon which the mujahedeen enjoy—after the help and granting of success by God—is popular support from the Muslim masses in Iraq, and the surrounding Muslim countries. So, we must maintain this support as best we can, and we should strive

---

[k] Of course, it is important to recognize that AQ was not merely being compassionate with its advice to not attack Shiites. Rather, AQ worried about igniting a sectarian war that could draw in Iran, the region's only Shiite power. Hoping to avoid this confrontation, and to win by means of popular support from Iraqi civilians, AQ leaders repeatedly cautioned restraint and were repeatedly rebuffed by ISIS leaders.

to increase it.... But you and your brothers must strive to have around you circles of support, assistance, and cooperation, and through them, to advance until you become a consensus, entity, organization, or association that represents all the honorable people and the loyal folks in Iraq. I repeat the warning against separating from the masses, whatever the danger.[80]

Zawahiri likely knew that in direct confrontations with US and Iraqi state forces, AQ and ISIS by themselves stood little chance. But if they could win the allegiance of the majority of Iraqi Sunnis, they might prevail.

While Zawahiri and others repeatedly advised ISIS leaders to spare Sunnis, these warnings fell on deaf ears. Zarqawi, Al-Muhajir, and both Al-Baghdadis rarely strayed from their own strategic visions. Underscoring how important this disagreement was, ISIS's disdain for Iraq's Sunni community is a key reason why the Sunni Awakening was so successful—when tens of thousands of Sunni tribesmen withdrew their support for insurgents and began collaborating with the Iraqi government.

Third, and finally, AQ was angered by ISIS's killing of hostages and criminals, and by its frequent public beheadings. As with other points of contention between them, these actions ran entirely counter to AQ's population-centric strategy; while they might terrify some inhabitants into submission, they might also erode the voluntary support of others. This division is reflected in some of the correspondence between leaders of AQ and IS. In one, Zawahiri writes that "Among the things which the feelings of the Muslim populace who love and support you will never find palatable... are the scenes of slaughtering the hostages." And in another letter, he writes:

... We are in a battle, and that more than half of this battle is taking place in the battlefield of the media. And that we are in a media battle in a race for the hearts and minds of our Umma.[1] And that however far our capabilities reach, they will never be equal to one thousandth of the capabilities of the kingdom of Satan that is waging war on us. And we can kill the captives by bullet. That would achieve that which is sought after without exposing ourselves to the questions and answering to doubts. We don't need this.[81]

As usual, this advice had little effect. But ISIS's brutality was starting to have a tangible, negative effect on Al-Qaeda. As Bacon and Arsenault write,[82] "AQI's

---

[1] Umma means "nation" in Arabic. In this context it can be understood as community of believers.

violence was so damaging to Al-Qaeda that the former head of the Bin Laden unit at the Central Intelligence Agency, Michael Scheuer, characterized Al-Zarqawi as 'the most potent strategic threat Al-Qaeda faced after 9/11.'" This left AQ in an untenable position.

Taken together, these strategic and ideological disagreements created a separate problem: one of control and of respect. Most evidence suggest that the leaders of AQI and its various incarnations had pledged bayat to Bin Laden or Zawahiri, as I discussed earlier. Yet, they behaved as if they were equal and blatantly ignored their advice and even commands. This meant that in addition to debates over strategy, targeting, and ideology, leaders of AQ frequently implored Zarqawi and others simply to consult them before making important decisions. In 2005, for instance, Atiyah Abd Al-Rahman asked "that you abstain from making any decision on a comprehensive issue (one with a broad reach), and on substantial matters until you have turned to your leadership; Shaykh Usamah [Bin Laden] and the Doctor [al-Zawahiri], and their brothers there, and consulted with them."[83] This was Atiyah's first recommendation on a long list. AQ leaders later sent similar letters to Abu Hamza Al-Muhajir as well.[84]

These issues of leadership and respect were of no small matter to AQ. Their brand and indeed their success partly depended upon their status as the vanguard of the jihadist movement. If ISIS were seen to ignore them, it could wear away at this carefully crafted image. It should therefore come as no surprise that after their split, ISIS took aim directly at this perception, putting out a statement describing how "al Qaeda today has ceased to be the base of jihad."[85] While issues of control and respect are common among armed groups, few have played out over so long and few have had such ramifications for the entire jihadist movement.

Notwithstanding these issues, the proverbial straw that broke the camel's back—the trigger for the split—was Baghdadi's unilateral, unsanctioned merger with JN, and ISIS's broader expansion into Syria. This was, fundamentally, a dispute over strategy: Zawahiri wanted the Islamic State to remain in Iraq, whereas Baghdadi wanted to conduct operations in neighboring Syria and to take JN under its wing.

Evidence that this event triggered the split comes from two sources. First, when compared to the disagreements described above, this was perhaps the only new event that temporally preceded the rift. The ideological disagreements, the slaughter of Muslims, the disobedience—all had been taking place for years while AQ and ISIS managed to cooperate. ISIS's actions in Syria, however, had only begun about a year prior to their rupture. Second, one can understand the seriousness of Zarqawi's sentiments from the letter he wrote to the leaders of ISIS and JN. When the merger occurred in Spring 2013, he wrote that:

> Sheikh Abou Bakr Al-Baghdadi was wrong when he announced the Islamic State in Iraq and the Levant without asking permission or receiving advice from us and even without notifying us. Sheikh Abu Mohamed Al-Joulani was wrong by announcing his rejection to the Islamic State in Iraq and the Levant, and by showing his links to Al-Qaeda without having our permission or advice, even without notifying us. The Islamic State in Iraq and the Levant is to be dissolved, while Islamic State in Iraq is to continue its work. Jabhat Al-Nusra is an independent entity for Qaedat Al-Jihad group, under the (al-Qaeda) general command. The seat of the Islamic State in Iraq is in Iraq. The seat of Jabhat Al-Nusra for the people of Al-Sham is in Syria.[86]

This was a clear rebuke centered entirely on ISIS actions vis-à-vis JN and Syria. Not only did it nullify the merger, but it even dissolved ISIS (referred to here as the Islamic State in Iraq and the Levant) and limited their operations to Iraq. Few public communications mirror this tone and directness, underscoring how serious of an issue it was.

Ultimately, ISIS's unique trajectory makes it difficult to neatly categorize this event as an organizational rupture and to clearly identify its causes. The group that Bin Laden brought into AQ (JTJ) already had a distinct strategy, ideology, and cult of personality built around Zarqawi. This would suggest a multidimensional fracture. But the triggering event—the one that seemed to truly be too much—was overwhelmingly strategic and unidimensional in nature. Consequently, the split exhibits characteristics of both: the AQI/ISI/ISIS faction was distinct along many dimensions, but it was strategic differences that overwhelmingly caused the schism.

Depicting the Split in ISIS's *Dabiq* Magazine

Another way to understand the split between ISIS and AQ is to examine the groups' public communications. These documents, like the Islamic State's *Dabiq* magazine, are useful because they reveal armed groups' personal attempts to shape the public narrative about the split and to portray themselves vis-à-vis their former partner. These narratives are important because they influence how people understand each organization, and these perceptions can affect people's decisions about which group to support or even which to join.

In this section, I conduct a content analysis of ISIS's *Dabiq* magazine.[m] There are 14 issues in total ranging from July 2014 through April 2016, and the first was produced in the same year as IS's split with AQ. The magazine was initially released monthly then later became more infrequent. IS delivered *Dabiq* online, through the deep web, and through various other channels.

Of the fourteen issues I examine, only three mention Al-Qaeda: issues #1 (2014), #2 (2014), and #6 (2016). This fact alone suggests that ISIS's magazine tends to focus more on its own organization and its unilateral successes, and less on the dynamics of conflict in Iraq and Syria or on its relationships with other groups.

In issue #1 Al-Qaeda is mentioned twice on page 33 in a section titled "The Islamic State in the Words of the Enemy." This part of the magazine is dedicated to an article published by two American researchers—Douglas A. Ollivant and Brian Fishman—and prominently features a picture of Ollivant at a Cato Institute event. The quote describes ISIS as "no longer a state in name only. It is a physical, if extra-legal, reality on the ground." And it refers to ISIS as the "former Al-Qaeda affiliate" that now "holds territory, provides limited services, dispenses a form of justice (loosely defined), most definitely has an army, and flies its own flag."[87] It is interesting that ISIS chose to publicize an analysis from two Americans focused on their state-like status.

The second quote that mentions Al-Qaeda is from the same page (33) and section ("The Islamic State in the Words of the Enemy"). It reads:

> Finally, this new reality presents a challenge that rises above a mere counter-terrorism problem. ISIS no longer exists in small cells that can be neutralized by missiles or small groups of commandos. It is now a real, if nascent and unrecognized, state actor—more akin in organization and power to the Taliban of the late 1990s than Al-Qaeda.

---

[m] ISIS issues an English-language version of *Dabiq*, and I consulted these English-language editions through the *Jihadology* website run by Aaron Y. Zelin: https://jihadology.net.

As with the previous quote, this once again highlights ISIS's achievements. However, it also draws a comparison with other groups, arguing that ISIS's success makes it more like the Taliban—an official state government through the 1990s—than Al-Qaeda. This obviously reflects well on a group with stated goals of becoming a caliphate. It is also noteworthy that this analysis portrays IS as more successful than Al-Qaeda in meeting its goals, but it is not overly critical of Al-Qaeda either.

Taken together, the first issue of *Dabiq* provides little information about ISIS's split, and it is careful not to denigrate its former collaborator. Nonetheless, ISIS intentionally publicizes research that draws a stark comparison between the two groups and that highlights ISIS's success in developing into a state-like authority, which AQ was never able to do.

The second issue of *Dabiq* is a different matter. IS authors directly broach the topic of intra-jihadi tensions in a section titled "The Flood of Mubahalah." This section is devoted to the conflicts between IS and other groups in the region, particularly JN. It castigates JN leadership for "slandering" Abu Umar Al-Baghdadi and Abu Hamzah Al-Muhajir. It notes that "these two men [al-Baghdadi and Al-Muhajir] and the state that they established were praised by Shaykh Usamah Ibn Ladin (rahimahullah)[n] as well as the rest of the former Al-Qa'idah leadership."[88] It is interesting that it refers to the "former" AQ leadership, not mentioning Zawahiri and others, and that it specifically notes the approval from Osama Bin Laden, even though he was killed nearly three years prior. This underscores the reverence they, and surely many of their potential recruits, still held for him.

The second issue of *Dabiq* again mentions Al-Qaeda on page 26:

In contrast, despite what the Islamic State faces of economic, military, political, and media war, and despite all the different parties unified against it—from the new Al-Qa'idah leadership in Khurasan, to the [S]afawis in Tehran, and all the way to the crusaders in Washington—it advances from victory to victory.... It killed rafidah ("Muslims" according to the new Al-Qa'idah leadership) by the thousands.[o] It kept to its promise and destroyed the border obstacles that formerly separated the lands of Iraq from Sham. Its numbers continue to grow.

[n] In Arabic, this translates to "God have mercy on him."
[o] Rafidah literally translates to "rejecters" in Arabic, and in this context it refers to those not accepting Islam (the Sunni faith in particular) and to those not sufficiently devoted.

This quote is notable because it explicitly defines Al-Qaeda as one of the "different parties unified against it," lumping it together with the United States and Iran. And, neither this issue of *Dabiq* nor the previous one make any mention that ISIS used to be part of AQ. This approach follows on the trend identified earlier, noting ISIS's comparative success but not denigrating AQ any further. This type of analysis also aligns with the conclusion drawn by Gambhir, who says that "This framing fits with ISIS's overarching critique of AQ: that the organization has neglected its duty to work toward the establishment of a Caliphate."[89] Apart from that, however, ISIS has little interest in discussing them.

The final mention of AQ is in issue #6 published half a year later. This is also the last time ISIS mentions their former collaborator in its magazine. However, it is not IS authors writing about AQ this time; instead, *Dabiq* reproduces the testimony of an Al-Qaeda defector in Waziristan, one of the tribal areas of northeast Pakistan. The testimony is full of blistering critiques, though a few stand out as they underscore some of the more important differences between the two groups. For instance:

> [Zawahiri] then continually appeared in the media displaying himself to be a gentle lamb on the issue of the State and the bay'ah and insisted upon having "the right" for leadership and "the obligation" to be listened to and obeyed. He began to describe the State and its leader as having the worst traits....
>
> [Zawahiri's] ideas contradict jihad and the carrying of arms, and encourage pacifist methodologies [never-ending protests] and the seeking of popular support, all of which led to the new Pharaohs' takeover of Egypt and other countries. Many women, children, and men were killed for no reason. They could not do anything except go to the roads, squares, and plazas, practicing the new politics that [Zawahiri] called to and those like him who claimed that what the protestors practice is the real jihad which will change oppression into justice and kufr into Islam.... He thereby destroyed Tandhim Al-Qa'idah.

This passage is particularly interesting because it alludes to core differences in strategy between AQ and IS—a fundamental reason for their split. However, instead of portraying AQ as more moderate but nonetheless violent, AQ is described as actually being pacifist. This approach, according to *Dabiq*, led to many protesters being killed in the Arab Spring, and then to the destruction of Al-Qaeda. The author also notes that in addition to

AQ's pacifism, Zawahiri repeatedly lied about IS activities. "They fabricated lies against us and described us in the harshest of ways: takfiiri, Khawarij, killers of Muslims, Wahhabi. . . and they would warn the people that we were murderers and that we'd slaughter them."[90] Though the critique is not wrong, ISIS apparently did not appreciate being called out in this way.

Aside from this, much of the defector's published testimony aims to personally discredit Zawahiri's stewardship of AQ. The author claims Zawahiri was "leading [AQ] to the bottom of the pit." And as for AQ, it was "a drowning entity struggling to breathe in deep water as it is exhausted and fatigued by tiredness and the struggle in the water." He claimed they were also cutting off monetary support for the families of fighters[91] and generally misappropriating donors' funds.[92] These passages account for the most direct critiques of AQ so far, but interestingly, they come from a defector and not directly from ISIS or its leaders.

Taken together, the messaging strategy in *Dabiq* makes sense in some ways but is curious in others. It makes sense that IS would want to distance itself from AQ, to establish its own reputation, and to play up its hardline credentials. These are typical strategies for breakaway groups that need to establish their reputation vis-à-vis their parent organization. Yet, I would also expect splinter groups to utilize this platform to go beyond establishing their own reputation and to attack that of their former partner. ISIS only does this in a meaningful way in its sixth issue and not even in its own words. This decision could be due to the fact that ISIS is already well-established, especially where it operates in Syria and Iraq, so there is no critical need to bash its rivals. Yet it contrasts with ISIS's behavior on the ground where physical confrontations with rivals were common.[93] In this regard, it is important to consider how *Dabiq* is aimed at foreign fighters and potential recruits from abroad.[94] ISIS might believe these individuals are less interested in their disagreements with AQ—disagreements that might be more important to local recruits who may know members of AQ and who may be targeted by AQ recruiters. Accordingly, playing up these disagreements and disparaging AQ could turn away some foreign fighters who have little interest in these intra-jihadi disputes.

Finally, it is noteworthy that there is no mention of the ideological fissures that divide AQ and ISIS. Of course, researchers like Gambhir[95] argue that "ISIS [does take] great care to ensure that its religious justification is robust," and that "*Dabiq* was released primarily as a tool to justify religious authority." But the magazine does *not* focus on ideological differences with AQ or others, which is something that theories of terrorist behavior might suggest.[96] Instead, any mention of AQ is in the context of a favorable strategic, tactical, or organizational comparison. As before, this approach may reflect an understanding of the population they are seeking to recruit: foreigners who may be less concerned with religious debates and more with religious devotion, notoriety, adventure, and victory.[97]

Lessons from the Islamic State's Split with Al-Qaeda

There is much for researchers and policymakers to learn from the development of ISIS, and my theory of how armed groups break down contributes to our understanding of it.

First, a close-knit, ideologically and strategically distinct group known as Jama'at Tawhid wal-Jihad developed in Iraq following the US invasion in 2003, initially around Abu Musab Al-Zarqawi and then around his successors. By 2004, it had officially affiliated itself with Osama Bin Laden and became Al-Qaeda in Iraq. Once merged, AQI operated much like a faction (as I define in earlier chapters) within AQ. It had a fundamentally different ethos underpinned most prominently by radical strategic differences. This had meaningful consequences. The fighters Zarqawi attracted were specifically drawn to the organizational niche he filled: a niche, or what we may think of as a brand, that fused dogmatic religious views with brutal violence. As Koerner writes, this positioning "allowed the Islamic State to rouse followers that Al-Qaeda never was able to reach."[98] These followers were ruthless, they were unwavering, and they were willing to carry out Zarqawi's brutal strategic vision. And since Zarqawi and his successors had almost no connections in Iraq, this recruitment pattern cannot be explained by community ties or social networks. Instead, it shows how individuals join armed groups and factions within them owing to their objective affinity. This underscores how the goals and attributes of armed groups influence their recruitment, which subsequently influences their internal, organizational dynamics as well.

Second, upon exiting Al-Qaeda and becoming fully independent, ISIS went on to fulfill ideological and strategic goals that were long evident to many observers. Its brutal violence, its strict enforcement of Sharia law, its establishment of an Islamic caliphate, and its attacks on Sunnis and even more brutal attacks on Shiites can all be traced back to how the group operated while it was still part of AQ, and even before. There is no "mystery of ISIS," as an anonymous author in the *New York Review of Books* notes—but perhaps only a mystery to those who were not paying attention.[99] As I make clear in my theory of how armed groups divide, evidence of how breakaway groups will behave is visible long before they become independent. The same is true of the Islamic state.

Third, the split between ISIS and Al-Qaeda reveals how researchers must take the disagreements within armed groups seriously. While such disputes are ubiquitous, we can see from this case that they are not fleeting or insignificant, they are not necessarily products of state intervention or a lack of control, and they are also not inconsequential to future patterns of group and intergroup behavior. Instead, the disagreements described in this chapter reflected long-standing tensions that fundamentally shaped the trajectories of both organizations.

Fourth, we could not have anticipated the exact timing of ISIS's eventual break with Al-Qaeda, but we should have known that one was coming. Although it was ISIS's attempt to commandeer Jabhat Al-Nusra that ultimately ended its relationship with Al-Qaeda, fundamental strategic and ideological differences punctured their cooperation for nearly a decade. At any point, Zawahiri could have decided to sever their relationship. Relatedly, analysts would have been hard pressed to anticipate the event that mattered most—that Abu Bakr Al-Baghdadi would claim JN for the Islamic State— or that this would finally push Ayman Al-Zawahiri to the breaking point. Instead, analysts could observe the significant tension between AQ and IS that raised the odds of a schism, but any more precision than that would be difficult to establish. This underscores the need for combining real-time monitoring with case experts who are deeply familiar with the inner workings of specific armed groups.[P]

Fourth, just as with many other splinter groups, ISIS's leaders had wisely taken steps to prepare for their eventual organizational independence. Zarqawi was most prescient in this matter. He established the Mujahideen Shura Council that helped unify the disparate, competitive landscape of Iraqi insurgents. This would prove especially useful when IS declared a state in territory where many of these groups had stronger local ties.[100] ISIS also established its own supply lines in and out of Iraq so that they were not entirely dependent on AQ's benevolence.[101] When war broke out in Syria, it took advantage of the chaos to expand and solidify its operational base. All of these actions were important if ISIS were to thrive in a world where it was no longer under the AQ umbrella. The fact that so many of these

---

[P] This event also reveals what quantitative analyses of organizational fractures might overlook.

actions were taken prior to their split suggests that ISIS was actively and intentionally preparing for such a scenario. In future research, it would be useful to investigate the success of breakaway groups when they do and not take precautions such as these.

Fifth, and also in line with my theoretical expectations, there was no lack of infighting within the Islamic State once it became independent. As with the Irish National Liberation Army, disagreements emerged because of the group's relatively broad organizational niche. Some members joined to fight US forces and to liberate Iraq, others for the mere opportunity of violence, while others still joined to fulfill religious imperatives. This disagreement was most obvious between former members of the Iraqi army, whose goals were mainly nationalist in nature, and foreign fighters who were often highly devout. According to Harris, "Baathists want the ouster of [Iraqi prime minister Nouri Al-]Maliki, to regain some of the stature and political participation that they've been denied since the fall of Saddam Hussein.[102] And that's a very different goal from setting up a caliphate." Many of the Baathists "[viewed] the jihadists with this Leninist mind-set that they're useful idiots who we can use to rise to power."[103] While these tensions were never fully resolved, they can be linked to ISIS's diverse objectives and to the multifaceted nature of its differences with AQ. But why did these disagreements not lead to another organizational rupture or more simply to organizational decline as it did with many other groups? One possible explanation lies in ISIS's strict oversight of its members and its harsh punishment of rulebreakers. Beatings, lashings, and stonings were not uncommon, both for members of the public and even for group members. As I allude to earlier, these are potential, though costly, solutions to a divided membership base,[104] and ISIS's ruthless adherence to this strategy may explain its efficacy.

Sixth, the disagreements between armed groups can also shed light on the odds they eventually cooperate or engage in direct confrontations with another. Though not my primary focus, I anticipate that some disputes—like those over strategy—are indeed surmountable. While it was ultimately strategy that prompted Zawahiri to expel ISIS, there were also ideological disagreements and some over control and respect. None of these, however, prevented nearly a decade of cooperation between AQ and ISIS, so it may be possible for them to collaborate again in the future, especially if leadership changes occur. Of course, the two groups have battled one another directly in Yemen,[105] Syria,[106] Somalia,[107] and elsewhere. But in West Africa, "Groups linked to Al-Qaeda and the Islamic State . . . are working together to take control of territory." According to Brigadier General Dagvin Anderson, head of the US military's Special Operations arm in Africa, "What we've seen is not just random acts of violence under a terrorist banner but a deliberate campaign that is trying to bring these various groups under a common cause. That larger effort poses a threat to the United States."[108] It remains to be seen if these isolated acts of cooperation will lead to more widespread collaboration, but it is a dangerous possibility. Their history of successful cooperation in the past may remind them of what benefits still remain, especially as one or both groups become weakened.

Seventh, the merger between JTJ and AQ reveals that actions designed to increase cooperation between armed groups—just as those designed to increase security, or to increase control—are not cost-free. Instead, they require each of the organizations involved to make important tradeoffs. In this case, while AQ gained a foothold into Iraq and some additional capacity, it introduced significant tension into its ranks by essentially on-boarding a distinct faction that it would struggle to control. And, for its part, JTJ lost some autonomy and became tied to AQ's successes and failures. Clearly, these costs were unevenly distributed. But since this was not the only means of cooperation, it is worth considering the risks of different arrangements in future research and why groups might select into one or another. So far, much of this variation has been overlooked.

Finally, as I have stressed elsewhere, it is important to recognize that the breakup of AQ and ISIS was not a clear-cut case of organizational splintering for two reasons. To begin with, it is dubious whether a full organizational merger that effectively created a single group ever took place. The Islamic State, and AQI before that, maintained a meaningful degree of organizational autonomy, had a fundamentally different outlook, and in some ways operated independently. Personnel, finances, and materiel were certainly shared, but they were not unified. If this were a true merger, one would expect to see greater control, coordination, overlap, and a clearer (and more respected) chain of command. In addition, this case was not a clear organizational split since it was AQ that expelled ISIS. In most schisms, it is the dissenting faction that chooses, voluntarily, to exit the organization. ISIS was not a case of a dissenting faction developing, solidifying, and then breaking away. Instead, it was more like a medical transplant that worked for some time but was eventually rejected by the host. Little is known about organizational mergers, including why they succeed or not, and future research should tackle this important topic.

6

Conclusions, Implications, and Future Research

Overview and Main Findings

This book explains why splinter groups—organizations that break away from preexisting armed groups to become independent—radicalize and survive to different extents. While splintering is common, researchers have largely ignored the intraorganizational politics that terminate in a splinter's emergence. In this book, I refocus attention towards the disagreements that initially lead subgroups to form and then break away, and I show how these disagreements subsequently shape their future trajectories. By understanding who joins splinter groups and what they hope to accomplish, we can more readily anticipate how they will eventually behave.

This book develops my argument in several stages. In Chapter 1, I show how subnational conflicts are becoming more fragmented with more non-state actors getting involved. This phenomenon is partly explained by the splintering of existing armed groups, and numerous data sets (including my own) reveal that around 30 percent of all armed groups form in this way. This underscores how important it is to understand the dynamics of group breakdown and splinter formation. Yet, existing research mostly black-boxes this process and instead focuses on the external or organizational factors associated with organizational splits, such as repressive state intervention and hierarchical leadership structures. These studies overlook internal political dynamics that shape and influence group breakdown in important ways. Similar shortcomings permeate discussions about fragmentation as a counterinsurgency strategy. Those advocating for it typically focus on how their interventions can rupture militant organizations while giving little thought to how it may accelerate their decline and the conflict's resolution. Underlying this an assumption that fragmented groups are weaker or more short lived, but as I show here, that is far from universal.

Divided Not Conquered: How Rebels Fracture and Splinters Behave. Evan Perkoski, Oxford University Press.
© Oxford University Press 2022. DOI: 10.1093/oso/9780197627075.003.0006

In Chapter 2, I develop a theory to explain variation in rates of survival and radicalization among groups emerging from schisms in militant organizations. I argue that the number of disagreements motivating splinter groups has implications for survival and cohesion, while the types of disagreements shape the extent to which they radicalize. Regarding the number of disagreements, groups forming unidimensionally (over a single, shared dispute with their parent organizations) benefit from consistent, aligned internal preferences. This leaves them more resilient since they are able to avoid feuds and decentralize their operations. Regarding the types of disagreements, groups forming over strategic disputes tend to attract a larger number of tactical and strategic hardliners who ultimately make the organization more likely to pursue a radical path. Consequently, the behavior of splinter groups is neither random nor fixed, but it depends on the disputes motivating their formation.

In Chapter 3, I present a comparative case study of two Irish Republican splinter groups: the Irish National Liberation Army (INLA) and the Real Irish Republican Army (RIRA). These groups make for an ideal comparison as they share a host of similarities but differ in their reasons for forming. The INLA formed multidimensionally around strategic and ideological disagreements with its parent group, the Official IRA. As a result, members drawn to the INLA had varying, often conflictual aspirations that led to both verbal and physical altercations. Internal feuds ultimately tore the organization apart and prevented it from decentralizing its command structure. Furthermore, the INLA's centralization of command inspired even more feuding as operatives coveted the top leadership spot. The RIRA, on the other hand, is an exemplary unidimensional fracture based on strategic disagreements. The RIRA's predecessor, the Provisional IRA, engaged in ceasefires and compromises that angered Republican hardliners who sought to achieve change immediately through violence. These hardliners migrated to the RIRA and they were dogmatic about the primacy of violent resistance. While the RIRA's homogeneous membership composition increased the group's resiliency and permitted decentralization, it also explains its unwavering commitment to violence. Unlike the INLA that perpetually debated the relative merits of violent and political strategies, the RIRA was steadfast in its violent approach. Ultimately, this case study demonstrates how the membership composition of splinter groups is shaped by the disagreements motivating them to form and subsequently drives variation in their survival and radicalization.

In Chapter 4, I provide a robust statistical analysis of my theory. I begin by discussing the methodology behind my new data set of organizational fractures among 300 randomly selected armed groups. Analyzing this data, I find that all types of groups are prone to internal ruptures, but the longer a group exists, the more likely they become. I then use this new data to test my prediction that splinter groups forming unidimensionally will survive longer. Using survival analyses and controlling for a battery of covariates that might also influence rates of endurance, I find strong evidence in support of my argument. Splinter groups as a whole are neither more nor less likely to survive, but they exhibit patterns of failure similar to the average nonsplinter group. Groups forming unidimensionally, however, greatly outlast both multidimensional splinters and even nonsplinter groups. Controlling for other factors, groups from unidimensional splits are nearly 25% more likely to survive than those forming multidimensionally. I then disaggregate unidimensional schisms to further examine this trend. While groups that splinter for personal and ideological reasons survive and fail at rates similar to nonsplinters, militant groups emerging from strategic schisms tend to endure the longest. I theorize this is because strategic schisms produce organizations with closely aligned preferences over their future actions. In addition, their commitment to violence may lead to a level of dedication which exceeds that in other kinds of splinter organizations. Ideological and personal disputes, meanwhile, produce groups that are not necessarily united around behavior in any meaningful way, leaving them at a slight disadvantage. Nonetheless, each type of unidimensional splinter is expected to outlast groups from multidimensional ruptures, as I expect.

Next, I use this data to test my prediction that splinter groups emerging from strategic disagreements are the most likely to radicalize. I theorize that these fractures produce the most violent breakaway groups since they typically attract tactical and strategic hardliners from their parent, and they drive the group's decision-making toward the extreme. Variation in yearly attack frequency, casualties, and per-attack casualty rates support my intuition. These differences are especially apparent in a group's first year of activity when they may be seeking to establish their reputation, but it holds throughout their entire lifespans as well. Among these different metrics, one of the most consistent findings is that strategically motivated splinters kill many more people per individual attack. This trend holds for groups that form unidimensionally over issues of strategy, but also when strategy is one issue among several others. Taken together, this is strong evidence that the reasons leading breakaway groups to form are meaningfully linked to their short- and long-term violent behavior.

In Chapter 5, I study the formation of the Islamic State (IS). While not a typical case of organizational splintering, my framework still helps make sense of the group's evolution. In effect, Al-Qaeda (AQ) imported a faction into their organization when in 2004 they merged with Jama'at Tawhid Wal-Jihad, Abu Musab Al-Zarqawi's organization that had recently relocated from Afghanistan to Iraq. Zarqawi's organization, now called Al-Qaeda in Iraq (AQI), is the progenitor of IS. While disagreements constantly plagued AQ's relationship with AQI, it was a final strategic dispute in 2014 that led them to part ways. The internal politics that preceded this rupture help us understand not only the split, but how IS would evolve on its own. IS's brutal tactics and dogmatic ideological views did not crystallize after their split, but they actually contributed to it. The act of becoming independent was largely inconsequential in this regard. If outside observers had recognized this, they could have foreseen the rupture and anticipated the challenge that IS would pose both to stability in Iraq and to AQ's global reputation. That knowledge could have been used by coalition forces to exacerbate their weak relationship earlier on, to hasten their uncoupling, and then to isolate and attack the separate entities. The international community's failure to take seriously these intraorganizational politics is a key reason why it was blindsided by IS's emergence and swift rise.

Policy Implications

The findings from this book have ramifications for policymakers and practitioners tasked with defeating armed groups and preventing the rise of new challengers to peace and stability.

From a counterterrorism and counterinsurgency perspective, my conclusions suggest that policymakers should proceed with caution. "Divide and conquer" strategies may be attractive since breaking up a nonstate foe can always be framed as a political victory,[1] but doing so is inherently unpredictable. The groups that emerge when an organization breaks apart are not universally weaker, less radicalized, or more short-lived. Instead, fragmentation can reshuffle group members and produce smaller, more numerous, and sometimes even more cohesive organizations, but it does little to address the motivations underlying the conflict or the people who fight in it. In comparison to the logic of repression (which aims to dissuade individuals from fighting by raising the cost of participation) or conciliation (which aims to dissuade fighters by giving in to their demands), there is no clear logic underlying the divide and conquer approach. If armed groups participated in more direct combat operations, then perhaps such divisions would reduce their military capacity. Except for a handful of outliers, however, they generally do not. So while having fewer members in one united group may prevent large-scale operations, it has little effect on the types of attacks most commonly launched by terrorists and insurgents.

My research also has implications for states pursuing more conciliatory approaches to violent groups. Specifically, negotiations with militant organizations that already suffer from internal divisions can be dangerous. Compromising or negotiating with armed groups where factions already exist can upend their internal balance and hasten organizational ruptures. Negotiations can exacerbate divisions between moderates and hardliners, compelling individuals to coalesce into subgroups that support or oppose a potential deal. When those who favor violence and reject compromise break away, the result can be a hardline, cohesive spoiler that poses serious challenges to ongoing negotiations and to lasting peace.[2] Consequently, governments should carefully consider if their militant adversaries are in a position to remain united during negotiations. If not, then negotiations may not be worth the risk.

What can governments do to prevent fragmentation? Since factions are strategic and rational when deciding to break away, states can take actions to affect their strategic calculus by demonstrating the futility of their departure. To this end, states can signal their intent to negotiate exclusively with one group. This could make splinters less likely to form if they believe they cannot can get a better deal on their own. Similarly, states can prevent the rise of some dissident challengers by demonstrating their commitment to fight any new armed groups that emerge, and by maintaining their military presence even after a deal takes place. If potential splinters believe they can break away and operate amid a more lenient environment during or after a negotiated troop withdrawal, they may decide to take their chances. But if a state can credibly signal its resolve, potential splinters may be deterred. This logic was applied to US negotiations with the Afghan Taliban; amid the backdrop of an inevitable US withdrawal, hardline, Taliban members may have rejected American offers since they knew US forces would soon be leaving the country either way.[3]

States contemplating negotiations with an armed nonstate actor should therefore carefully consider the timing of any negotiations. They should avoid negotiating too early in a conflict since a significant portion of an armed group may still believe that military success is possible. Under such circumstances, a settlement offer could produce a large, motivated breakaway group intent on pursuing violence. If, however, a militant group's morale is low and its odds of success seem slim, potential spoilers may be deterred by the futility of their prospects. While these findings offer little in the way of precise guidance, they do suggest that states should take steps to gauge the morale of enemy combatants and to use this as part of their negotiation strategy.

Of course, there can also be scenarios where fragmentation may be the only option for states seeking a comprehensive settlement with an armed adversary. This might occur when a state sees no chance of improving its bargaining position and faces domestic pressure to withdraw or to compromise. Here, a policy designed to foment a strategic dispute among insurgents could be one way to separate hardliners and moderates, facilitating negotiations with moderates while enabling the state to concentrate its resources on hardliners. One astute Congressional Research Service report identified the advantages of this strategy, noting that "A psychologically sophisticated policy of promoting divisions between political and military leaders as well as defections within guerrilla and terrorist groups is likely to be more effective than a simple military strategy based on the assumption that all members and leaders of the group are hard-liners."[4] Following this advice, states could pursue a counterinsurgency strategy that uses conciliatory offers to divide a group, winning over moderates and leaving only hardliners behind.[5] This sort of approach depends upon states carefully considering the dynamics within armed groups and being committed to fighting new groups that emerge. For instance, states seeking to divide moderates from hardliners might consider leaking or openly embracing the details of a potential settlement to help identify dissenting factions. This approach could also allow the leaders of armed groups to take preventive actions to consolidate their control by proactively expelling or sanctioning dissenters. While this option certainly poses risks, it can lead to the disarmament of the majority of fighters—as happened in the Northern Ireland peace agreement of 1998. The Provisional IRA disarmed and entered into political negotiations with the United Kingdom, while a smaller, but more manageable, subset of militants (the Real IRA) continued to fight. Even though a relatively small group can still pose a meaningful threat, it may come in the form of sporadic acts of violence that a state is willing to tolerate for the sake of progress.

Fomenting organizational fractures can potentially be effective in other circumstances, too, especially when they yield new groups that are short-lived and plagued by internal disagreements of their own. To achieve this outcome, states should aim to create multidimensional schisms within militant groups. They should target them with a variety of simultaneous interventions—everything from playing one leader against another, to limited conciliation, to propaganda operations that discredit ideological and strategic goals—in order to cause widespread, cross-cutting discord. This multifaceted approach has the greatest chance of producing organizational ruptures that reduce levels of violence and that hasten organizational demise. Conversely, states should avoid policies that are singularly repressive or singularly conciliatory. These may be highly counterproductive.[6] As one report from the US Federal Research Division (Library of Congress) even notes, "A military response to terrorism unaccompanied by political countermeasures is likely to promote cohesion within the group."[7]

Of course, it is critical to keep in mind that there is no *assured* way for states to foment a particular type of fracture within armed groups. We can neither reliably predict how events will reverberate among group members nor how factions will respond to different interventions. Even a multifaceted counterterrorist campaign, as described above, could just as likely compel factions to ramp up violence as it could inspire greater cohesion while rebels face down a much stronger adversary. There is an inherent unpredictability to social dynamics like these, and policymakers would be wise to account for uncertainty in their planning efforts.

Research Implications

In addition to the policy implications, the findings from this book are also relevant to the academic study of armed groups, and specifically, how they break down and ultimately function. To the first point, this is one of the first works to clearly delineate the process of organizational splintering and to isolate the inflection points that allow for variation to occur. In doing so, it sheds much-needed light on a highly common organizational process. To the second point, my work reveals how armed groups' observable behavior masks numerous less-observable processes operating below the surface, and how these processes can be traced to the manner in which groups initially formed. I show how the membership composition of splinter groups is partly determined when a faction coalesces within its parent and fundamentally affects their decision-making and behavior once independent. This underscores how armed groups are not merely products of their strategic environment, but are instead dynamic entities that are simultaneously shaped by both internal and external forces.

This book also underscores what a growing body of literature already suggests: that commonplace organizational dynamics are critically important to understanding how armed groups behave. These are not purely rational, unitary automatons,[a] but they are groups of people whose relationships and interactions are highly impactful. Especially in the case of splinter groups, these personal interactions and group dynamics shape the trajectory of the organization for years to come. Recognizing the connection between basic human nature and the behavior of armed groups is imperative as research progresses.

As to the topic of how armed groups form, existing research largely focuses on the incipient resource endowments, social connections, and ideologies of new organizations.[8] There is also a robust literature on armed groups that emerge from political parties, and more recently, from other types of organizations as well.[9] But, as I show here, it is not only from what types of antecedent organizations splinters emerge that is important, but it is just as valuable to understand why they emerged and whether it was cooperative or conflictual, or maybe even intentional. Researchers can learn much by applying this logic to armed groups that break off from all sorts of organizations.

To illustrate the value of the approach presented in this book, consider how it applies to a topic of significant interest: tactical diffusion. There is already much research on why groups adopt certain tactics like suicide bombings, improvised explosive devices, aerial hijackings, and so on.[10] These analyses treat armed groups as independent units, but many are not: they are often militant wings of larger organizations or splinter groups that broke away from existing groups. And these splinter groups sometimes band together others. A good example of this is Tehrik-i-Taliban Pakistan (TTP),

---

[a] Though, this does not preclude one from using rational choice frameworks to study group behavior. Depending on the level of analysis and on the particular research question, this is still a highly valuable perspective, but it should take into account limitations imposed by human dynamics.

which is essentially an umbrella organization for formerly independent armed groups operating in Pakistan's tribal region.[11] TTP did not "adopt" suicide terror in the same way that the Tamil Tigers of Sri Lanka did, which sought out the capability and had to learn it from other organizations. Some of the TTP's founding members already possessed the needed expertise. This allowed them to implement the tactic immediately. As such, the experiences of groups like the TTP obscure empirical findings related to tactical diffusion and potentially misrepresent the influence of interorganizational cooperation and other factors. The same is true for the growing body of research on the life-cycles of armed groups and how they evolve as they mature.[12] Life-cycle dynamics should be significantly different for breakaway groups and others that do not form from the ground up. Their formative years may not be as difficult (or they may be difficult for different reasons) and they may reach maturity more quickly. Their organizational needs may therefore be fundamentally different. Since many armed groups form by splintering, it is imperative to take this information into account.

Similar to research findings on organizational life-cycles, I also find that splinter groups behave somewhat differently in their first year as independent entities. Splinter groups tend to be more active than nonsplinter groups upon forming, and those emerging from strategic disputes are especially prone to violence. Since disagreements over strategy commonly occur around government negotiations, researchers should consider how levels of violence may fluctuate before, immediately after, and then long after agreements take place. While I postulate that this first-year escalation may stem from a group's desire to establish its identity and hardline credentials, other organizations may make similar adjustments to their behavior as they mature. Groups emerging from political parties may face analogous motivations. While there is some research on the age of armed groups, especially from the perspective of tactical adoption, my findings suggest that it may be linked to many characteristics of group behavior.

Although my findings are most directly relevant to explaining the organizational dynamics of specific armed groups, they have implications for understanding conflict-level dynamics as well. As I demonstrate in Chapter 1, conflicts around the globe are growing increasingly fragmented and many groups have emerged through processes of organizational splintering. While it appears that around 30 percent of all armed groups form in this way, fragmentation in places like Iraq and Syria may even be more prevalent.[13] The theoretical framework I present here can be useful in such scenarios. Knowing if groups are splinters and from which parent organization they emerged can help researchers more effectively explain and anticipate patterns of competition and cooperation. In contrast to existing approaches that assume similar strategies based on aggregate conflict statistics (e.g., a higher number of active organizations will be linked to increased competition), data on group lineages could offer more specific predictions for dyadic relations and more nuanced insights in general. Likewise, tracking the emergence of new groups in a conflict, as well as their reasons for breaking away, would shed light on the types of contentious issues permeating the opposition. Patterns of strategic disagreement could even portend conflict escalation as hardline offshoots emerge, whereas leadership disputes might signal growing intra-group competition and an absence of order or hierarchy. The types of disagreements might even evolve as conflicts mature. In effect, accounting for why armed groups are dividing can offer new insights into fragmented conflicts.

Potential Applications and Future Research

My research reveals a number of important questions worth exploring in future work. First, my theory of splinter formation and behavior is not necessarily limited to violent nonstate actors. On the contrary, it is likely that fractures among other types of organizations will follow similar trajectories, going from informal faction to independent splinter, and understanding their causes can be instructive. For instance, the US civil rights movement experienced a split "between organizations that maintained that nonviolent methods were the only way to achieve civil rights goals and those organizations that had become frustrated and were ready to adopt violence and black separatism."[14] This is remarkably similar to the types of disputes commonly plaguing armed groups. At the same time, civil resistance groups do not solely disagree over the use of force. In another noteworthy event, a leader of the US Students for a Democratic Society "[declared] that it was impossible to remain in the same organization with people who opposed self-determination in practice and demanded an immediate split."[15] Researchers and policy-makers can leverage the disagreements within nonviolent groups to better understand their cohesion and their potential adherence to a nonviolent strategy. And, as I discuss in Chapter 2, organizational splits may even be in a group's best interest if the alternative is a nonviolent organization remaining unified with members engaging in unsanctioned violence. Indeed, violence attributed to the organization could negate some of the core benefits of nonviolent resistance, so it may be useful to separate out factions that fundamentally disagree over matters of strategy.[16]

Second, while this project focuses on the radicalization and survival of splinter groups, my theoretical framework can be extended to explain other outcomes. It is likely that the odds of conflict between parent organizations and splinter offspring are related to characteristics of their fracture. Different disagreements should correspond with different likelihoods that parents and splinters either compete with or fight against one another. Notably, personal feuds might be the most likely to inspire direct fighting as leaders hold grudges against their former partners and against those who did not side with them during a split. On the other hand, ideological splits may produce new groups that share similar, though not identical, beliefs. They could be most likely to outbid for local support as they are essentially competing over the same pool of recruits and exhibit few differences in strategy. And finally, cooperation might be eventually possible when armed groups divide over strategic issues since they still share the same goals while only disagreeing over how to achieve them. Though hard to imagine when their split took place in 2014, even Al-Qaeda and ISIS have overcome some of their strategic differences and are now cooperating, albeit on a limited basis, in some parts of the world.[17] The possibility of reunification between splinters and parents is troubling, especially in this case, and more research on this subject is needed.

Third, future research should investigate what happens to parents groups after a split. To be sure, there is already some work on this particular topic. On the one hand, researchers argue that parent groups are at risk of collapsing when larger factions depart.[18] On the other hand, splits can actually benefit armed groups when they pull away dissenting individuals and help create more aligned internal preferences among the remaining fighters.[19] My argument offers a potential link between these findings: parent groups are most likely to reap benefits from unidimensional splits, where breakaway groups with focused grievances attract a specific subset of dissenting members like anti-compromise hardliners or religious purists.[b] Because of their smaller niche, these groups will tend to pull away fewer operatives. Splinter groups with large niches, however, will attract members from various segments of the group, diminishing the potential benefits while simply robbing their predecessor of more members. Nonetheless, future research should examine the relationship between parent and progeny organizations, and how the former survives a challenge to its leadership.

Fourth, my research—especially the case studies—underscores the need for more work on the role of particular individuals within rebel organizations.[20] Specifically, I find that certain individuals—such as Michael McKevitt in the RIRA, or Eduardo Moreno "Pertur" Bergareche in ETA—were critical not only to the eventual trajectory of their breakaway groups, but to the initial formation of factions within their parent organizations. This raises an obvious counter-factual: would these splits have happened without their involvement? While I evaluate this possibility and find that these ruptures probably would occurred without them, these examples reveal the critical role that influential leaders and members can play. More work could be done, for instance, on how cults of personality shape internal dynamics and decision-making, and this could have long-term consequences for armed groups.

Fifth, the findings from this book suggest that researchers should examine armed groups that emerge from other sources like political parties and national militaries. Political parties can spawn armed groups unintentionally when a dissenting set of hardliners breaks away, but also more cooperatively when members recognize a strategic value to a simultaneous armed campaign. These two pathways will result in very different relationships between the political party and armed group, and only in the latter (cooperative) model will the armed group reap the organizing benefits of the parallel political apparatus. Relatedly, my research makes it clear that coalitional dynamics and members' discourse both have an impact on how armed groups behave. Researchers should therefore investigate similarities in decision-making between armed groups and other entities. By comparing them to political parties, which more explicitly rely on and are shaped by coalitional dynamics, we can better understand how armed organizations function.

---

[b] This will also be beneficial for states since separating out ideological or strategic hardliners allows them to tailor their strategies. This potentially helps to explain why fragmented conflicts experience more concessions by the state (e.g. Cunningham, 2011).

With regard to discourse, it plays an essential role in determining how factions form, grow, and eventually exit their parent group. Since discursive choices, such as how to frame disagreements or which disagreements to publicize, can have a meaningful impact on group dynamics, discourse analysis provides a promising avenue for future studies.

Appendix

Primary Statistical Results with Full List
of Covariates and Estimates

Table A.1 Full results: Splinter and nonsplinter violence in year one

	Attacks	Fatalities	Average Lethality	Suicide Bombing
Splinter Group	1.033	1.974**	1.428	3.860**
	(0.149)	(0.636)	(0.417)	(2.650)
Observations	287	287	287	287

Standard errors in parentheses (clustered by group). Year fixed effects included.
Results reported as incident rate ratios. * $p<.1$, ** $p<.05$, *** $p<.01$

Table A.2 Full results: Strategic splinter violence in year one

	Attacks	Fatalities	Average Lethality	Suicide Bombing
Splinter, Strategy	0.990	2.763***	2.242**	2.442
Involved	(0.165)	(1.004)	(0.767)	(2.158)
Splinter, No Strategy	1.125	0.825	0.501*	6.300**
	(0.289)	(0.419)	(0.194)	(5.002)
Observations	287	287	287	287

Standard errors in parentheses (clustered by group).
Year fixed effects included.
* $p<.1$, ** $p<.05$, *** $p<.01$

Table A.3 Full results: Disaggregating splinter violence in year one

	Attacks	Fatalities	Average Lethality	Suicide Bombing
Strategic Splinter	1.004	3.127***	2.243**	1.694
	(0.195)	(1.352)	(0.915)	(1.925)
Ideological Splinter	3.596**	4.034**	1.299	8.750*
	(1.821)	(2.460)	(0.769)	(10.451)
Leadership Splinter	0.618	0.084***	0.145***	1.000
	(0.329)	(0.044)	(0.069)	(.)
Multidimensional	0.718*	0.694	0.677	4.200
Splinter	(0.123)	(0.289)	(0.299)	(3.751)
Observations	286	286	286	280

Standard errors in parentheses (clustered by group).

Year fixed effects included.

Results reported as incident rate ratios. * $p<.1$, ** $p<.05$, *** $p<.01$

Table A.4 Full results: Splinter and nonsplinter violence

	Attacks	Fatalities	Average Lethality
Splinter Group	0.694**	0.778	1.177
	(0.121)	(0.149)	(0.139)
Group Age	1.030**	1.018	0.993
	(0.015)	(0.018)	(0.010)
Nationalist-Separatist	1.120	0.823	0.772**
	(0.209)	(0.191)	(0.096)
Communist-Socialist	1.408	0.725	0.553***
	(0.358)	(0.218)	(0.098)
Religious	1.501*	1.273	0.960
	(0.320)	(0.388)	(0.130)
Leftist	1.040	0.453	0.446***
	(0.371)	(0.283)	(0.138)
Transnational	3.626***	4.277***	1.190*
	(0.488)	(0.632)	(0.121)
Autocracy	0.587**	0.753	1.500
	(0.137)	(0.228)	(0.384)
Democracy	1.291	1.261	0.981
	(0.338)	(0.306)	(0.153)
Regime Durability	1.008	0.997	0.987***
	(0.008)	(0.007)	(0.005)
GDP Per Capita (log)	0.879	0.718	0.515***
	(0.245)	(0.253)	(0.121)

	Attacks	Fatalities	Average Lethality
Competitors	1.027***	1.022**	0.996
	(0.010)	(0.011)	(0.008)
Organizational Alliances	1.054	1.098	1.056***
	(0.059)	(0.068)	(0.009)
Observations	1877	1877	1877

Standard errors in parentheses (clustered by group).

Country and year fixed effects included.

* p<.1, ** p<.05, *** p<.01

Table A.5 Full results: Strategic splinter violence

	Attacks	Fatalities	Average Lethality
Splinter, Strategy Involved	0.735	0.849	1.288**
	(0.149)	(0.185)	(0.166)
Splinter, No Strategy	0.564**	0.536*	0.838
	(0.127)	(0.176)	(0.179)
Group Age	1.029**	1.017	0.993
	(0.015)	(0.019)	(0.010)
Nationalist-Separatist	1.116	0.814	0.763**
	(0.209)	(0.193)	(0.096)
Communist-Socialist	1.396	0.726	0.556***
	(0.354)	(0.215)	(0.098)
Religious	1.529**	1.315	0.994
	(0.330)	(0.401)	(0.140)
Leftist	1.032	0.453	0.442***
	(0.367)	(0.281)	(0.137)
Transnational	3.624***	4.232***	1.179
	(0.490)	(0.631)	(0.120)
Autocracy	0.580**	0.732	1.438
	(0.137)	(0.225)	(0.369)
Democracy	1.333	1.331	0.998
	(0.356)	(0.336)	(0.154)
Regime Durability	1.009	0.997	0.987***
	(0.008)	(0.008)	(0.005)
GDP Per Capita (log)	0.867	0.684	0.510***
	(0.243)	(0.248)	(0.121)
Competitors	1.027***	1.022**	0.996
	(0.010)	(0.011)	(0.008)
Organizational Alliances	1.053	1.098	1.056***
	(0.059)	(0.068)	(0.009)
Observations	1877	1877	1877

Standard errors in parentheses (clustered by group).

Country and year fixed effects included.

* p<.1, ** p<.05, *** p<.01

Table A.6 Full results: Disaggregating splinter violence

	Attacks	Fatalities	Average Lethality
Strategic Splinter	0.809	0.992	1.440**
	(0.155)	(0.221)	(0.214)
Ideological Splinter	0.887	1.233	1.561
	(0.287)	(0.517)	(0.515)
Leadership Splinter	0.262***	0.150***	0.554
	(0.090)	(0.098)	(0.297)
Multidimensional Splinter	0.499**	0.372***	0.695*
	(0.165)	(0.143)	(0.138)
Group Age	1.033**	1.024	0.998
	(0.015)	(0.019)	(0.010)
Nationalist-Separatist	1.048	0.737	0.711***
	(0.200)	(0.179)	(0.092)
Communist-Socialist	1.350	0.679	0.509***
	(0.347)	(0.216)	(0.094)
Religious	1.506**	1.317	0.969
	(0.314)	(0.389)	(0.134)
Leftist	0.987	0.396	0.413***
	(0.342)	(0.242)	(0.127)
Transnational	3.601***	4.250***	1.185
	(0.475)	(0.610)	(0.122)
Autocracy	0.597**	0.769	1.543*
	(0.143)	(0.232)	(0.401)
Democracy	1.293	1.252	0.991
	(0.346)	(0.324)	(0.156)
Regime Durability	1.008	0.996	0.986***
	(0.008)	(0.007)	(0.005)
GDP Per Capita (log)	0.894	0.753	0.530***
	(0.249)	(0.270)	(0.126)
Competitors	1.027***	1.023**	0.996
	(0.010)	(0.011)	(0.008)
Organizational Alliances	1.047	1.080	1.052***
	(0.059)	(0.064)	(0.009)
Observations	1876	1876	1876

Standard errors in parentheses (clustered by group).

Country and year fixed effects included.

* $p<.1$, ** $p<.05$, *** $p<.01$

Evaluating the Impact of State Repression on Rates of Survival

Table A.7 and Table A.8 replicates the models of militant survival from Chapter 3 while adding additional covariates that measure the level of domestic repression. These variables come from the CIRI Human Rights Dataset (version 2014.04.14) available at http://www.humanrightsdata.com.

Table A.7 Cox proportional hazards model of organizational longevity: evaluating the effects of state repression

	Model 1	Model 2	Model 3	Model 4	Model 5
Unidimensional Splinter	0.546**	0.546**	0.544**	0.548**	0.563*
	(0.145)	(0.145)	(0.145)	(0.145)	(0.168)
Multidimensional Splinter	1.565	1.569	1.562	1.575	1.596
	(0.458)	(0.462)	(0.455)	(0.474)	(0.485)
Nationalist-Separatist	0.911	0.924	0.930	0.906	0.886
	(0.170)	(0.173)	(0.173)	(0.170)	(0.182)
Communist-Socialist	0.542**	0.555**	0.557**	0.549**	0.517**
	(0.149)	(0.151)	(0.151)	(0.151)	(0.158)
Religious	0.908	0.910	0.908	0.917	1.126
	(0.214)	(0.214)	(0.213)	(0.215)	(0.268)
Leftist	3.037***	2.790***	2.788***	2.802***	3.724***
	(0.803)	(0.743)	(0.742)	(0.742)	(1.206)
Autocracy	0.646	0.573	0.570	0.661	0.907
	(0.368)	(0.326)	(0.316)	(0.370)	(0.563)
Democracy	1.344	1.316	1.270	1.480	1.465
	(0.508)	(0.488)	(0.465)	(0.584)	(0.705)
Regime Durability	0.992	0.991	0.991	0.988	0.982
	(0.011)	(0.011)	(0.011)	(0.011)	(0.013)
GDP Per Capita (log)	0.634	0.719	0.675	0.907	0.886
	(0.258)	(0.298)	(0.280)	(0.400)	(0.519)
Population (log)	1.599**	1.555**	1.577**	1.547**	1.573**
	(0.325)	(0.317)	(0.321)	(0.332)	(0.359)
Terrorist Competitors	1.014	1.013	1.015	1.004	0.998
	(0.019)	(0.019)	(0.019)	(0.020)	(0.021)
Lead Organization	0.926	0.917	0.914	0.913	0.915
	(0.168)	(0.166)	(0.166)	(0.164)	(0.178)
Disappearances	1.002*				
	(0.001)				
Extrajudicial Killings		1.003**			
		(0.001)			
Political Imprisonment			1.000		
			(0.001)		
Torture				1.013**	
				(0.006)	
Physical Integrity Score					0.900
					(0.068)
Observations	2361	2361	2361	2361	2180

Standard errors in parentheses (clustered by group). Coefficients reported as hazard ratios.
Country and year fixed effects included.
* $p<.1$, ** $p<.05$, *** $p<.01$

Table A.8 Cox proportional hazards model of organizational longevity: evaluating the effects of state repression

	Model 1	Model 2	Model 3	Model 4	Model 5
Strategic Splinter	0.517**	0.518**	0.516**	0.518**	0.578*
	(0.146)	(0.146)	(0.146)	(0.146)	(0.178)
Ideological Splinter	0.303	0.287	0.288	0.300	0.256
	(0.285)	(0.272)	(0.271)	(0.282)	(0.252)
Leadership Splinter	0.712	0.715	0.706	0.747	0.856
	(0.352)	(0.347)	(0.339)	(0.369)	(0.502)
Multidimensional Splinter	1.574	1.577	1.571	1.585	1.626
	(0.458)	(0.463)	(0.455)	(0.474)	(0.494)
Nationalist-Separatist	0.909	0.921	0.926	0.906	0.879
	(0.170)	(0.173)	(0.173)	(0.170)	(0.179)
Communist-Socialist	0.530**	0.542**	0.544**	0.537**	0.496**
	(0.148)	(0.151)	(0.150)	(0.150)	(0.156)
Religious	0.940	0.946	0.944	0.950	1.181
	(0.229)	(0.230)	(0.229)	(0.231)	(0.291)
Leftist	2.885***	2.650***	2.654***	2.651***	3.347***
	(0.815)	(0.757)	(0.757)	(0.753)	(1.140)
Autocracy	0.575	0.507	0.505	0.579	0.764
	(0.366)	(0.323)	(0.314)	(0.364)	(0.538)
Democracy	1.323	1.298	1.250	1.447	1.452
	(0.505)	(0.487)	(0.464)	(0.575)	(0.710)
Regime Durability	0.993	0.992	0.993	0.990	0.984
	(0.012)	(0.012)	(0.012)	(0.011)	(0.013)
GDP Per Capita (log)	0.618	0.698	0.655	0.868	0.853
	(0.255)	(0.293)	(0.275)	(0.390)	(0.506)
Population (log)	1.612**	1.570**	1.592**	1.557**	1.586**
	(0.331)	(0.324)	(0.327)	(0.335)	(0.363)
Terrorist Competitors	1.013	1.012	1.014	1.004	0.999
	(0.019)	(0.019)	(0.020)	(0.020)	(0.021)
Lead Organization	0.930	0.920	0.918	0.917	0.917
	(0.170)	(0.167)	(0.167)	(0.166)	(0.180)
Disappearances	1.002				
	(0.001)				
Extrajudicial Killings		1.003**			
		(0.001)			
Political Imprisonment			1.000		
			(0.001)		
Torture				1.012**	
				(0.006)	
Physical Integrity Score					0.895
					(0.068)
Observations	2360	2360	2360	2360	2180

Standard errors in parentheses (clustered by group). Coefficients reported as hazard ratios.
Country and year fixed effects included.

* $p<.1$, ** $p<.05$, *** $p<.01$

As one can see, almost all estimates are unchanged, though the significance levels for multidimensional ruptures drops to just above p=.10 with the significantly fewer observations under analysis (due to limitations of the CIRI data). Some of the repression variables are, as one would anticipate, associated with a slight increase in the likelihood of group failure. Specifically, the prevalence of extrajudicial killings and torture. These variables are omitted from the main results since they reduce the number of observations.

Evaluating the impact of group lethality on rates of survival

Table A.9 replicates my models of militant survival while adding an additional covariate that measures the number of fatalities a militant group causes in a given year. This is done to investigate whether the intensity of the conflict might somehow influence my theoretical mechanisms. For instance, it is plausible that in conflicts where militants are causing more fatalities in a given year—a proxy for high-intensity conflicts—they are more likely to persevere despite high levels of internal preference divergence. The nature of the conflict and the existential threats posed to the organization might substitute for cohesion, keeping the organization together.

The results in Table A.9 demonstrate that my theory—and in particular how it relates to the survival of militant organizations—seems to function regardless of the conflict's intensity. The estimates of both unidimensional and multidimensional splinters, as well as the disaggregated typology, retain roughly the same effect size and the same significance level as before. Interestingly, the estimate of yearly fatalities is also significant with a moderating effect; in other words, groups that manage to kill greater numbers of individuals in a given year are more likely to survive. Although the precise mechanism between the two is unclear, it could be that this measure is proxying for group capacity—in which case more capable groups are less likely to fail in a given year—or it could be how I originally theorized: that higher intensity conflicts are associated with greater militant durability as these organizations bind together in the face of external threats.

Table A.9 Cox proportional hazards model of organizational longevity: evaluating the effects of group lethality

	Model 1	Model 2
Unidimensional Splinter	0.510***	
	(0.118)	
Multidimensional Splinter	1.487*	
	(0.354)	
Strategic Splinter		0.480***
		(0.120)
Ideological Splinter		0.333
		(0.251)
Leadership Splinter		0.650
		(0.292)

Continued

Table A.9 *Continued*

	Model 1	Model 2
Multidimensional Splinter		1.485*
		(0.354)
Nationalist-Separatist	0.880	0.876
	(0.146)	(0.145)
Communist-Socialist	0.646**	0.635**
	(0.141)	(0.141)
Religious	0.934	0.954
	(0.209)	(0.219)
Leftist	2.269***	2.237***
	(0.553)	(0.556)
Autocracy	0.459	0.437
	(0.238)	(0.240)
Democracy	1.045	1.024
	(0.302)	(0.298)
Regime Durability	0.990	0.990
	(0.009)	(0.010)
GDP Per Capita (log)	0.811	0.793
	(0.309)	(0.305)
Population (log)	1.685**	1.691**
	(0.376)	(0.376)
Terrorist Competitors	1.020	1.018
	(0.016)	(0.017)
Lead Organization	1.377*	1.385*
	(0.236)	(0.238)
Yearly Fatalities	0.984***	0.984***
	(0.006)	(0.006)
Observations	2910	2909

Standard errors in parentheses (clustered by group). Coefficients reported as hazard ratios.
Country and year fixed effects included.
* $p<.1$, ** $p<.05$, *** $p<.01$

Testing the proportionality assumption

Figure A.1, Figure A.2, Figure A.3, and Figure A.4 plot post-tests of the proportionality assumption that is crucial to using the Cox proportional hazards model. I test for proportionality across each of the different types of splinter organizations. If proportional, the residuals should be clustered around zero. As each graph shows, the mean residual is indeed zero and this holds across time. This strongly supports the use of Cox model to analyze the duration of militant organizations, and it also suggests that organizational formation has a constant, enduring effect on rates of cohesion and collapse.

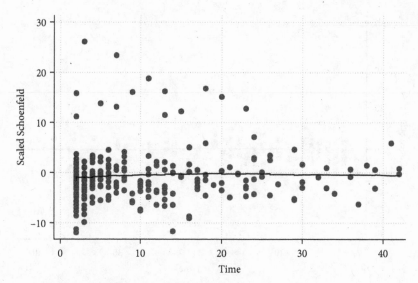

Fig. A.1 Test of proportionality: strategic splinters.

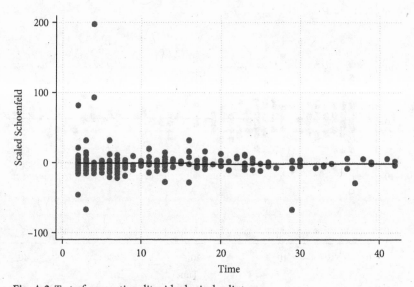

Fig. A.2 Test of proportionality: ideological splinters.

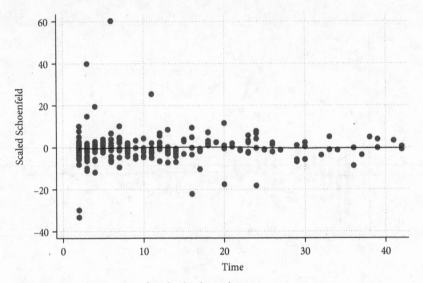

Fig. A.3 Test of proportionality: leadership splinters.

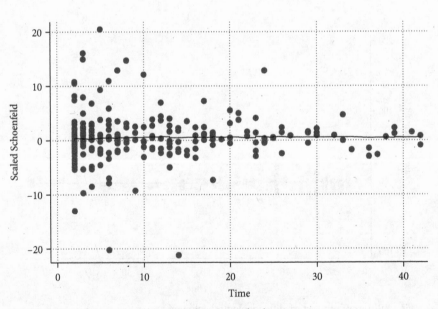

Fig. A.4 Test of proportionality: multidimensional splinters.

Evaluating rates of radicalization with ordinary least squares

Tables A.10, A.11, and A.12 replicate each of the models for group radicalization using ordinary least squares regression instead of negative binomial, and the results are similarly clustered by group. The same patterns emerge in terms of coefficient size and direction, but statistical significance weakens significantly. This is to be expected when moving from a model that is specifically designed for non-negative counts (negative binomial regression) to one designed for anything along the real number scale, positive or negative.

Table A.10 Splinter and nonsplinter violence with ordinary least squares

	Attacks	Fatalities	Average Lethality
Splinter Group	−2.710	−0.704	0.607
	(5.935)	(15.988)	(0.698)
Group Age	0.434	−0.180	−0.164*
	(0.459)	(1.254)	(0.089)
Nationalist-Separatist	−2.034	−20.335	−1.713*
	(5.225)	(13.488)	(0.895)
Communist-Socialist	−0.223	−17.630	−0.769
	(7.717)	(22.843)	(1.250)
Religious	1.845	−2.196	0.161
	(4.178)	(12.181)	(0.860)
Leftist	−11.213	−31.026	−1.366
	(9.278)	(25.329)	(1.197)
Transnational	30.098***	77.179***	0.631
	(6.690)	(20.427)	(0.832)
Autocracy	−4.351	−3.130	2.519
	(5.794)	(16.892)	(1.808)
Democracy	14.195	20.352	0.516
	(13.740)	(40.710)	(0.956)
Regime Durability	0.157	0.085	0.022
	(0.402)	(1.045)	(0.053)
GDP Per Capita (log)	−9.953	−36.247	−8.115
	(9.902)	(37.808)	(6.123)
Competitors	0.319	0.502	−0.014
	(0.271)	(0.754)	(0.041)
Organizational Alliances	−0.227	3.478	0.850***
	(1.251)	(2.938)	(0.280)
Observations	1877	1877	1877

Standard errors in parentheses (clustered by group).

Country and year fixed effects included.

* $p<.1$, ** $p<.05$, *** $p<.01$

Table A.11 Strategic splinter violence with ordinary least squares

	Attacks	Fatalities	Average Lethality
Splinter, Strategy Involved	−1.551	4.194	0.808
	(7.151)	(18.899)	(0.814)
Splinter, No Strategy	−6.652	−17.354	−0.078
	(5.522)	(17.221)	(1.115)
Group Age	0.437	−0.165	−0.164*
	(0.460)	(1.254)	(0.090)
Nationalist-Separatist	−2.138	−20.773	−1.731*
	(5.319)	(13.672)	(0.895)
Communist-Socialist	−0.244	−17.719	−0.773
	(7.670)	(22.557)	(1.243)
Religious	2.262	−0.438	0.233
	(4.021)	(11.723)	(0.882)
Leftist	−11.392	−31.782	−1.397
	(9.363)	(25.522)	(1.192)
Transnational	30.001***	76.770***	0.614
	(6.676)	(20.362)	(0.831)
Autocracy	−4.621	−4.274	2.472
	(5.921)	(17.064)	(1.811)
Democracy	14.685	22.422	0.601
	(14.116)	(41.674)	(0.986)
Regime Durability	0.158	0.091	0.022
	(0.403)	(1.047)	(0.053)
GDP Per Capita (log)	−10.080	−36.782	−8.137
	(9.913)	(37.907)	(6.125)
Competitors	0.322	0.515	−0.013
	(0.272)	(0.758)	(0.041)
Organizational Alliances	−0.241	3.419	0.848***
	(1.253)	(2.937)	(0.280)
Observations	1877	1877	1877

Standard errors in parentheses (clustered by group).

Country and year fixed effects included.

* $p<.1$, ** $p<.05$, *** $p<.01$

Table A.12 Disaggregating splinter violence with ordinary least squares

	Attacks	Fatalities	Average Lethality
Strategic Splinter	−1.950	1.736	1.122
	(3.596)	(10.152)	(1.024)
Ideological Splinter	−5.811	−17.918	0.557
	(8.035)	(21.381)	(1.787)
Leadership Splinter	−20.415	−65.344	−0.981
	(20.760)	(70.374)	(3.624)

	Attacks	Fatalities	Average Lethality
Multidimensional Splinter	−0.917	9.135	−0.129
	(14.178)	(39.165)	(0.819)
Group Age	0.442	−0.174	−0.155*
	(0.459)	(1.255)	(0.094)
Nationalist-Separatist	−2.102	−20.058	−1.900**
	(5.010)	(13.611)	(0.904)
Communist-Socialist	−0.189	−17.111	−0.926
	(6.749)	(19.845)	(1.284)
Religious	2.622	0.866	0.206
	(4.132)	(12.019)	(0.889)
Leftist	−11.474	−31.917	−1.484
	(9.265)	(25.334)	(1.191)
Transnational	30.021***	76.890***	0.610
	(6.640)	(20.273)	(0.832)
Autocracy	−4.590	−4.466	2.522
	(6.107)	(17.560)	(1.837)
Democracy	14.388	21.282	0.570
	(14.015)	(41.441)	(1.017)
Regime Durability	0.158	0.096	0.020
	(0.411)	(1.067)	(0.054)
GDP Per Capita (log)	−10.263	−37.508	−8.121
	(9.979)	(38.005)	(6.157)
Competitors	0.332	0.567	−0.013
	(0.271)	(0.753)	(0.042)
Organizational Alliances	−0.275	3.315	0.842***
	(1.297)	(3.066)	(0.285)
Observations	1876	1876	1876

Standard errors in parentheses (clustered by group).

Country and year fixed effects included.

*p<.1, ** p<.05, *** p<.01

Analyzing Rates of Radicalization without Fixed Effects

Table A.13 replicates the main analyses for splinter radicalization while omitting the country and year fixed effects. Fixed effects offer a robust method to account for unexplained variation across countries and over time, but with the relatively small sample size here, they may be absorbing some group-level variation as well. When omitted, the results support my theory even more strongly. In fact, splinters from strategic disputes exhibit statistically higher rates of radicalization across all three metrics—yearly attacks, fatalities, and average lethality—and to a greater extent than before.

Table A.13 Disaggregating splinter violence without fixed effects

	Yearly Attacks	Yearly Fatalities	Yearly Avg. Lethality
Strategic Splinter	0.526***	0.614**	1.641**
	(0.121)	(0.151)	(0.331)
Ideological Splinter	1.199	0.916	1.072
	(0.552)	(0.371)	(0.310)
Leadership Splinter	0.246***	0.189**	0.507
	(0.063)	(0.132)	(0.306)
Multidimensional Splinter	2.456	1.399	0.720*
	(1.643)	(0.969)	(0.121)
Group Age	1.045***	1.035**	0.988
	(0.011)	(0.016)	(0.008)
Nationalist-Separatist	0.673*	0.567*	0.831
	(0.161)	(0.164)	(0.123)
Communist-Socialist	1.374	0.525	0.582***
	(0.428)	(0.216)	(0.102)
Religious	0.912	0.805	1.008
	(0.211)	(0.322)	(0.163)
Leftist	0.765	0.539	0.475***
	(0.288)	(0.241)	(0.120)
Transnational	3.694***	3.202***	1.043
	(0.636)	(0.430)	(0.121)
Autocracy	0.472**	0.755	1.751*
	(0.145)	(0.293)	(0.510)
Democracy	0.928	1.049	1.059
	(0.228)	(0.338)	(0.187)
Regime Durability	0.999	0.997	0.998
	(0.003)	(0.004)	(0.002)
GDP Per Capita (log)	0.767**	0.377***	0.496***
	(0.082)	(0.055)	(0.042)
Competitors	0.987	0.971**	0.990
	(0.009)	(0.012)	(0.007)
Organizational Alliances	1.026	1.135**	1.054***
	(0.044)	(0.060)	(0.014)
Observations	1910	1910	1910

Standard errors in parentheses (clustered by group).

Country and year fixed effects included.

* $p<.1$, ** $p<.05$, *** $p<.01$

Evaluating the Link between Parent Groups and Splinter Behavior

Table A.14 assesses the extent to which a parent group's behavior is correlated with a splinter's. It is plausible that the two are linked since splinters emerging from preexisting organizations carry with them tactical and strategic know-how but also expectations about the use of force (which they may wish to alter). I therefore rerun the main analyses

Table A.14 Splinter radicalization and parent behavior

	Attacks	Fatalities	Avg. Lethality
Strategic Splinter	5.989***	12.946***	5.029**
	(2.065)	(11.578)	(3.517)
Ideological Splinter	9.713***	9.898**	2.125
	(3.486)	(11.217)	(1.572)
Multidimensional Splinter	2.172***	5.205*	3.359*
	(0.616)	(4.815)	(2.373)
Group Age	1.054***	1.010	0.947***
	(0.017)	(0.028)	(0.014)
Nationalist-Separatist	0.289***	0.566	1.510
	(0.076)	(0.229)	(0.421)
Communist-Socialist	0.738	0.317***	0.412**
	(0.175)	(0.122)	(0.155)
Religious	2.084***	1.214	0.739
	(0.454)	(0.536)	(0.238)
Leftist	1.770	1.039	0.297***
	(0.627)	(0.532)	(0.115)
Transnational	2.896***	3.189***	0.845
	(0.401)	(0.730)	(0.212)
Autocracy	0.723	0.993	1.278
	(0.512)	(0.816)	(1.266)
Democracy	1.576	1.573	1.076
	(0.489)	(0.988)	(0.536)
Regime Durability	1.006*	0.998	0.995
	(0.003)	(0.006)	(0.005)
GDP Per Capita (log)	0.663***	0.478***	0.576***
	(0.071)	(0.108)	(0.112)
Competitors	1.007	1.005	0.992
	(0.010)	(0.016)	(0.016)
Parent, Attacks	1.000		
	(0.000)		
Parent, Fatalities		1.000	
		(0.000)	
Parent, Avg. Attack Lethality			0.985***
			(0.004)
Splinter Group			
Observations	714	714	714

Standard errors in parentheses (clustered by group).

Country and year fixed effects included.

* p<.1, ** p<.05, *** p<.01

presented earlier that evaluate splinter radicalization, but I now include data on parent groups: their cumulative attacks, number of fatalities, and their average lethality.[a] This significantly lowers the number of observations for two reasons: first, I am now only analyzing splinter groups; and, second, I do not have attack data on parent groups in about 20% of cases (some of these parents existed pre-1970 before GTD data is available, and other parents simply are not listed in GTD).[b] It is also important to note that the reference category cannot be nonsplinter groups, as it was before, so it is now splinters motivated by ideological disputes. I chose this since I have no theoretical expectations about their radicalization.

These results show that parents and splinters are statistically correlated in terms of their average attack lethality, and interestingly, that association is negative. Cutting against conventional wisdom, more lethal parents tend to produce less lethal splinters. Otherwise, the main results presented earlier still hold. In contrast to groups forming over leadership disputes, those motivated by strategy are immensely more radicalized across all three indicators: yearly attacks, yearly fatalities, and average yearly lethality. Taken together, these results underscore the strong, path-dependent effect of intraorganizational politics and disputes on the evolution of breakaway groups.

Description of Quantitative Data

Table A.15 contains a list of all armed groups under analysis and their manner of formation. This excludes 15 groups, from the initial list of 300, whose origins were unable to be conclusively determined. It was clear that some of these were splinters, however, and they were used to compute earlier descriptive statistics.

[a] I do not solely rely on statistics from the year prior to a split because that behavior could be anomalous. For instance, splinters could emerge during ceasefires where the parents are largely inactive, misrepresenting what information that possess

[b] In contrast to previous analyses, I also omit the variable capturing the number of alliances an armed group has in order to maximize observations even more. Substantively, this has no effect on the results.

Table A.15 Coding of splinter and nonsplinter groups

Group Name	Type 1	Type 2	Reason 1	Reason 2	Reason 3
2nd of June Movement	Splinter	Multidimensional	Ideology	Strategy	
Abu Nidal Organization (ANO)	Splinter	Unidimensional	Strategy		
Abu Sayyaf Group (ASG)	Splinter	Unidimensional	Strategy		
Acilciler	Splinter	Unidimensional	Ideology		
Al-Shabaab	Splinter	Unidimensional	Strategy		
Albanian National Army (ANA)	Splinter	Unidimensional	Strategy		
All Tripura Tiger Force (ATTF)	Splinter	Unidimensional	Strategy		
Ansar Allah	Splinter	Unidimensional	Personal		
Ansar Al-Dine (Mali)	Splinter	Unidimensional	Strategy		
Anti-Talks Faction	Splinter	Unidimensional	Strategy		
Armed Forces of National Resistance (FARN)	Splinter	Multidimensional	Ideology	Strategy	
Azawad National Liberation Movement (MNLA)	Splinter	Multidimensional	Ideology	Leadership	
Biswamohan Faction of National Liberation Front of Tripura (NLFT-BM)	Splinter	Multidimensional	Personal	Strategy	Ideology
Black September	Splinter	Unidimensional	Strategy		
Colonel Karuna Faction	Splinter	Multidimensional	Personal	Strategy	
Communist Party of Nepal—Maoist (CPN-M)	Splinter	Unidimensional	Strategy		
Continuity Irish Republican Army (CIRA)	Splinter	Unidimensional	Strategy		
Democratic Front for the Liberation of Palestine (DFLP)	Splinter	Unidimensional	Strategy		
Dev Sol	Splinter	Unidimensional	Strategy		
Dev Yol	Splinter	Unidimensional	Ideology		
Dima Halao Daoga (DHD)	Splinter	Unidimensional	Strategy		
ETA-*pm*	Splinter	Multidimensional	Strategy	Ideology	
Eritrean Peoples Liberation Front	Splinter	Unidimensional	Strategy		
Fatah Hawks	Splinter	Unidimensional	Strategy		
Fatah Uprising	Splinter	Unidimensional	Personal		

Continued

Table A.15 *Continued*

Group Name	Type 1	Type 2	Reason 1	Reason 2	Reason 3
Fatah Al-Islam	Splinter	Unidimensional	Personal		
Fuerzas Armadas Revolucionarias del Pueblo (FARP)	Splinter	Multidimensional	Ideology	Strategy	
God's Army	Splinter	Multidimensional	Ideology	Leadership	
Gulbuddin Hekmatyar group	Splinter	Unidimensional	Personal		
Harakat ul-Mujahidin (HuM)	Splinter	Multidimensional	Ideology	Strategy	
Hizballah	Splinter	Unidimensional	Strategy		
Irish National Liberation Army (INLA)	Splinter	Multidimensional	Strategy	Ideology	
Islamic Jihad Group (IJG)	Splinter	Multidimensional	Strategy	Personal	
Jaime Bateman Cayon Group (JBC)	Splinter	Unidimensional	Strategy		
Jaish-e-Mohammad (JeM)	Splinter	Multidimensional	Personal	Personal	
Jamiat ul-Mujahedin (JuM)	Splinter	Unidimensional	Personal		
Jammu and Kashmir Islamic Front	Splinter	Multidimensional	Strategy	Personal	
Janatantrik Terai Mukti Morcha (JTMM)	Splinter	Unidimensional	Ideology		
Janatantrik Terai Mukti Morcha—Goit (JTMM-G)	Splinter	Unidimensional	Ideology		
Japanese Red Army (JRA)	Splinter	Unidimensional	Strategy		
Jaya Krishna Goit faction	Splinter	Unidimensional	Personal		
Jemaah Islamiya (JI)	Splinter	Unidimensional	Ideology		
Kahane Chai	Splinter	Multidimensional	Strategy	Leadership	
Kangleipak Communist Party (KCP)	Splinter	Unidimensional	Strategy		
Karbi Longri National Liberation Front (KLNLF)	Splinter	Unidimensional	Strategy		
Karbi People's Liberation Tigers (KPLT)	Splinter	Unidimensional	Strategy		
Lashkar-e-Jhangvi	Splinter	Unidimensional	Strategy		
Lebanese Armed Revolutionary Faction (LARF)	Splinter	Multidimensional	Ideology	Leadership	
Loyalist Volunteer Forces (LVF)	Splinter	Unidimensional	Strategy		
May 15 Organization for the Liberation of Palestine	Splinter	Multidimensional	Ideology	Leadership	
Minni Minnawi faction	Splinter	Multidimensional	Personal	Strategy	

Group Name	Type 1	Type 2	Reason 1	Reason 2	Reason 3
Moro Islamic Liberation Front (MILF)	Splinter	Unidimensional	Ideology	Ideology	
Movement for Oneness and Jihad in West Africa (MUJAO)	Splinter	Multidimensional	Personal	Ideology	
Nestor Paz Zamora Commission (CNPZ)	Splinter	Multidimensional	Unknown		
Official Irish Republican Army (OIRA)	Splinter	Multidimensional	Strategy	Ideology	
Palestine Liberation Front (PLF)	Splinter	Unidimensional	Strategy		
Palestinian Islamic Jihad (PIJ)	Splinter	Unidimensional	Strategy		
Pedro Leon Arboleda (PLA)	Splinter	Multidimensional	Personal	Ideology	
Peykar	Splinter	Multidimensional	Ideology	Strategy	
Popular Front for the Liberation of Palestine, Gen Cmd (PFLP-GC)	Splinter	Unidimensional	Strategy		
Popular Resistance Committees	Splinter	Unidimensional	Strategy		
Popular Revolutionary Vanguard (VPR)	Splinter	Multidimensional	Ideology	Ideology	
Provisional Irish Republican Army	Splinter	Unidimensional	Strategy		
Purbo Banglar Communist Party	Splinter	Unidimensional	Strategy		
Real Irish Republican Army (RIRA)	Splinter	Unidimensional	Strategy		
Red Brigades Fighting Communist Party (BR-PCC)	Splinter	Multidimensional	Leadership	Ideology	
Red Brigades Fighting Communist Union (BR-UCC)	Splinter	Multidimensional	Leadership	Ideology	
Salafia Jihadia	Splinter	Unidimensional	Ideology		
Salafist Group for Preaching and Fighting (GSPC)	Splinter	Unidimensional	Strategy		
Shining Path (SL)	Splinter	Multidimensional	Ideology	Strategy	
Takfir wal-Hijra (Excommunication and Exodus)	Splinter	Multidimensional	Ideology	Strategy	
United People's Democratic Front (UPDF)—Bangladesh	Splinter	Unidimensional	Strategy		
1920 Revolution Brigades	Nonsplinter				
23rd of September Communist League	Nonsplinter				
28 May Armenian Organization	Nonsplinter				
Abdullah Azzam Brigades	Nonsplinter				
Abu Hafs Al-Masri Brigades	Nonsplinter				

Continued

Table A.15 *Continued*

Group Name	Type 1	Type 2	Reason 1	Reason 2	Reason 3
Action Directe	Nonsplinter				
Adan Abyan Islamic Army (AAIA)	Nonsplinter				
African National Congress (South Africa)	Nonsplinter				
Al-Borkan Liberation Organization	Nonsplinter				
Al-Faran	Nonsplinter				
Al-Zulfikar	Nonsplinter				
Al-Aqsa Martyrs Brigade	Nonsplinter				
Al-Arifeen	Nonsplinter				
Al-Mansoorian	Nonsplinter				
Al-Nawaz	Nonsplinter				
Al-Qaeda	Nonsplinter				
Al-Qaeda in Iraq	Nonsplinter				
Al-Qaeda in the Lands of the Islamic Maghreb (AQLIM)	Nonsplinter				
Al-Sa'iqa	Nonsplinter				
Al-quds Brigades	Nonsplinter				
Amal	Nonsplinter				
Ananda Marga	Nonsplinter				
Andres Castro United Front	Nonsplinter				
Ansar Al-Sunnah (Palestine)	Nonsplinter				
Ansar Al-Din	Nonsplinter				
Ansar Al-Islam	Nonsplinter				
Anti-American Arab Liberation Front	Nonsplinter				
Anti-Imperialist International Brigades	Nonsplinter				
Anti-terrorist Liberation Group (GAL)	Nonsplinter				
Arab Commando Cells	Nonsplinter				
Arab Communist Organization	Nonsplinter				

Group Name	Type 1	Type 2	Reason 1	Reason 2	Reason 3
Arab Liberation Front (ALF)	Nonsplinter				
Arab Struggle	Nonsplinter				
Arab Unionist Nationalist Organization	Nonsplinter				
Arabian Peninsula Freemen	Nonsplinter				
Argentine Anticommunist Alliance (AAA)	Nonsplinter				
Armata di Liberazione Naziunale (ALN)	Nonsplinter				
Armed Forces Revolutionary Council (AFRC)	Nonsplinter				
Armed Islamic Group (GIA)	Nonsplinter				
Armenian Revolutionary Army	Nonsplinter				
Armenian Secret Army for the Liberation of Armenia	Nonsplinter				
Army of God	Nonsplinter				
Asbat Al-Ansar	Nonsplinter				
Baader-Meinhof Group	Nonsplinter				
Babbar Khalsa International (BKI)	Nonsplinter				
Baloch Liberation Army (BLA)	Nonsplinter				
Bangladesh Communist Party	Nonsplinter				
Basque Fatherland and Freedom (ETA)	Nonsplinter				
Bersatu	Nonsplinter				
Black Hand	Nonsplinter				
Black Widows	Nonsplinter				
Bodo Liberation Tigers (BLT)	Nonsplinter				
Brigades of Iman Hassan-al-Basri	Nonsplinter				
Charles Martel Group	Nonsplinter				
Che Guevara Brigade	Nonsplinter				
Committee of Solidarity with Arab and Middle East Political Prisoners (CSPPA)	Nonsplinter				
Corsican National Liberation Front (FLNC)	Nonsplinter				

Continued

Table A.15 *Continued*

Group Name	Type 1	Type 2	Reason 1	Reason 2	Reason 3
Croatian Freedom Fighters	Nonsplinter				
Democratic Karen Buddhist Army (DKBA)	Nonsplinter				
Dukhta-ran-e-Millat	Nonsplinter				
East Turkistan Liberation Organization	Nonsplinter				
Egypt's Revolution	Nonsplinter				
Ejercito de Liberacion Nacional (Bolivia)	Nonsplinter				
Eritrean Liberation Front	Nonsplinter				
Farabundo Marti National Liberation Front (FMLN)	Nonsplinter				
February 28 Popular League (El Salvador)	Nonsplinter				
First of October Antifascist Resistance Group (GRAPO)	Nonsplinter				
Free Aceh Movement (GAM)	Nonsplinter				
Free Papua Movement (OPM-Organisasi Papua Merdeka)	Nonsplinter				
Front de Liberation du Quebec (FLQ)	Nonsplinter				
Front for the Liberation of Lebanon from Foreigners	Nonsplinter				
Generation of Arab Fury	Nonsplinter				
Great Eastern Islamic Raiders Front (IBDA-C)	Nonsplinter				
Greek Bulgarian Armenian Front	Nonsplinter				
Guerrilla Army of the Poor (EGP)	Nonsplinter				
Hamas (Islamic Resistance Movement)	Nonsplinter				
Hezb-e Wahdat-e Islami-yi Afghanistan	Nonsplinter				
Hizb-I-Islami	Nonsplinter				
Hizbul Mujahideen (HM)	Nonsplinter				
Hizbul Al-Islam (Somalia)	Nonsplinter				
Imam Hussein Brigade	Nonsplinter				
International Justice Group (Gama'a Al-Adela Al-Alamiya)	Nonsplinter				
Iraqi Liberation Army	Nonsplinter				
Irish Republican Army (IRA)	Nonsplinter				

Group Name	Type 1	Type 2	Reason 1	Reason 2	Reason 3
Isatabu Freedom Movement (IFM)	Nonsplinter				
Islamic Action Organization	Nonsplinter				
Islamic Army in Iraq (al-Jaish Al-Islami fi Al-Iraq)	Nonsplinter				
Islamic Courts Union (ICU)	Nonsplinter				
Islamic Jihad Brigades	Nonsplinter				
Islamic Liberation Organization	Nonsplinter				
Islamic Movement for Change	Nonsplinter				
Islamic Movement of Uzbekistan (IMU)	Nonsplinter				
Islamic Shashantantra Andolon (ISA)	Nonsplinter				
Jama'atul Mujahideen Bangladesh (JMB)	Nonsplinter				
Jenin Martyrs Brigades	Nonsplinter				
Jewish Defense League (JDL)	Nonsplinter				
Jewish Fighting Organization (Eyal)	Nonsplinter				
Jordanian Islamic Resistance	Nonsplinter				
Jund–Al–Sham for Tawhid and Jihad	Nonsplinter				
Jundallah	Nonsplinter				
Justice Commandos for the Armenian Genocide	Nonsplinter				
Kach	Nonsplinter				
Kamtapur Liberation Organization (KLO)	Nonsplinter				
Karen National Union	Nonsplinter				
Kata'ib Al-Khoul	Nonsplinter				
Khmer Rouge	Nonsplinter				
Kosovo Liberation Army (KLA)	Nonsplinter				
Ku Klux Klan	Nonsplinter				
Kuki National Liberation Front (KNLF)	Nonsplinter				
Kuki Revolutionary Army (KRA)	Nonsplinter				

Continued

Table A.15 *Continued*

Group Name	Type 1	Type 2	Reason 1	Reason 2	Reason 3
Kurdistan Workers' Party (PKK)	Nonsplinter				
Lashkar-e-Omar	Nonsplinter				
Lashkar-e-Taiba (LeT)	Nonsplinter				
Laskar Jihad	Nonsplinter				
Lebanese Liberation Front	Nonsplinter				
Lebanese National Resistance Front	Nonsplinter				
Lebanese Socialist Revolutionary Organization	Nonsplinter				
Liberation Battalion	Nonsplinter				
Liberation Tigers of Tamil Eelam (LTTE)	Nonsplinter				
Lord's Resistance Army (LRA)	Nonsplinter				
Lorenzo Zelaya Revolutionary Front (LZRF)	Nonsplinter				
M-19 (Movement of April 19)	Nonsplinter				
Manuel Rodriguez Patriotic Front (FPMR)	Nonsplinter				
Maoist Communist Center (MCC)	Nonsplinter				
Masada, Action and Defense Movement	Nonsplinter				
15-May	Nonsplinter				
Montoneros (Argentina)	Nonsplinter				
Moro National Liberation Front (MNLF)	Nonsplinter				
Movement for Democracy and Justice in Chad (MDJT)	Nonsplinter				
Movement for the Emancipation of the Niger Delta (MEND)	Nonsplinter				
Mozambique National Resistance Movement (MNR)	Nonsplinter				
Mujahedeen Army	Nonsplinter				
Mujahedeen Shura Council	Nonsplinter				
Mujahedin-e Khalq (MEK)	Nonsplinter				
Muslim Brotherhood	Nonsplinter				
Muslims Against Global Oppression (MAGO)	Nonsplinter				
Muttahida Qami Movement (MQM)	Nonsplinter				

Group Name	Type 1	Type 2	Reason 1	Reason 2	Reason 3
National Army for the Liberation of Uganda (NALU)	Nonsplinter				
National Democratic Front of Bodoland (NDFB)	Nonsplinter				
National Liberation Army (NLA) (Macedonia)	Nonsplinter				
National Liberation Army of Colombia (ELN)	Nonsplinter				
National Liberation Front of Tripura (NLFT)	Nonsplinter				
National Patriotic Front of Liberia (NPFL)	Nonsplinter				
National Union for the Total Independence of Angola (UNITA)	Nonsplinter				
Nepali Congress Party (NC)	Nonsplinter				
New People's Army (NPA)	Nonsplinter				
November 17 Revolutionary Organization (N17RO)	Nonsplinter				
Orly Organization	Nonsplinter				
Oromo Liberation Front	Nonsplinter				
Palestine Liberation Organization (PLO)	Nonsplinter				
Palestinian Revolution Forces	Nonsplinter				
Patriotic Morazanista Front (FPM)	Nonsplinter				
Pattani United Liberation Organization (PULO)	Nonsplinter				
Peace Conquerors	Nonsplinter				
Peasant Self-Defense Group (ACCU)	Nonsplinter				
People's Revolutionary Army (ERP)	Nonsplinter				
People's Revolutionary Army (ERP) (El Salvador)	Nonsplinter				
People's Revolutionary Party of Kangleipak (PREPAK)	Nonsplinter				
People's United Liberation Front (PULF)	Nonsplinter				
People's War Group (PWG)	Nonsplinter				
Peronist Armed Forces (FAP)	Nonsplinter				
Polisario Front	Nonsplinter				
Popular Front for the Liberation of Palestine (PFLP)	Nonsplinter				
Popular Liberation Army (EPL)	Nonsplinter				

Continued

Table A.15 *Continued*

Group Name	Type 1	Type 2	Reason 1	Reason 2	Reason 3
Popular Revolutionary Army (Mexico)	Nonsplinter				
Raul Sendic International Brigade	Nonsplinter				
Rebel Armed Forces of Guatemala (FAR)	Nonsplinter				
Red Army Faction (RAF)	Nonsplinter				
Red Brigades	Nonsplinter				
Red Hand Defenders (RHD)	Nonsplinter				
Revolutionary Action	Nonsplinter				
Revolutionary Armed Forces of Colombia (FARC)	Nonsplinter				
Revolutionary Liberation Action (Epanastatiki	Nonsplinter				
Apelevtherotiki Drasi) — Greece					
Revolutionary Organization of Socialist Moslems	Nonsplinter				
Revolutionary United Front (RUF)	Nonsplinter				
Salafi Abu-Bakr Al-Siddiq Army	Nonsplinter				
Sandinista National Liberation Front (FSLN)	Nonsplinter				
Save Kashmir Movement	Nonsplinter				
Secret Organization Zero	Nonsplinter				
Sipah-e-Sahaba/Pakistan (SSP)	Nonsplinter				
South Moluccans	Nonsplinter				
South-West Africa People's Organization (SWAPO)	Nonsplinter				
Spanish Basque Battalion (BBE) (rightist)	Nonsplinter				
Spanish National Action	Nonsplinter				
Students Islamic Movement of India (SIMI)	Nonsplinter				
Sudan Liberation Movement	Nonsplinter				
Sudan People's Liberation Army (SPLA)	Nonsplinter				
Syrian Social Nationalist Party	Nonsplinter				
Taliban	Nonsplinter				
Tawhid and Jihad	Nonsplinter				

Group Name	Type 1	Type 2	Reason 1	Reason 2	Reason 3
The Extraditables	Nonsplinter				
Tigray Peoples Liberation Front (TPLF)	Nonsplinter				
Tontons Macoutes	Nonsplinter				
Tripura National Volunteers (TNV)	Nonsplinter				
Tupac Amaru Revolutionary Movement (MRTA)	Nonsplinter				
Tupamaros (Uruguay)	Nonsplinter				
Turkish Communist Party/Marxist (TKP-ML)	Nonsplinter				
Turkish Islamic Jihad	Nonsplinter				
Turkish People's Liberation Front (TPLF)(THKP-C)	Nonsplinter				
Uganda Democratic Christian Army (UDCA)	Nonsplinter				
Ulster Freedom Fighters (UFF)	Nonsplinter				
Ulster Volunteer Force (UVF)	Nonsplinter				
United Liberation Front of Assam (ULFA)	Nonsplinter				
United National Liberation Front (UNLF)	Nonsplinter				
United People's Democratic Solidarity (UPDS)	Nonsplinter				
United Self Defense Units of Colombia (AUC)	Nonsplinter				
West Nile Bank Front (WNBF)	Nonsplinter				
White Legion (Ecuador)	Nonsplinter				
Workers' Revolutionary Party	Nonsplinter				
Zarate Willka Armed Forces of Liberation	Nonsplinter				
Zimbabwe African Nationalist Union (ZANU)	Nonsplinter				
al-Ahwaz Arab People's Democratic Front	Nonsplinter				
al-Fatah	Nonsplinter				
al-Gama'at Al-Islamiyya (IG)	Nonsplinter				
al-Intiqami Al-Pakistani	Nonsplinter				

Notes

Chapter 1

1. Claire M. Hart and Mark Van Vugt (2006). "From Fault Line to Group Fission: Understanding Membership Changes in Small Groups." *Personality and Social Psychology Bulletin* 32.3, pp. 392–404, p. 392.
2. John Horgan (2012). *Divided We Stand: The Strategy and Psychology of Ireland's Dissident Terrorists*. Oxford University Press, p. 21.
3. Yoram Schweitzer (2011). "Innovation in Terrorist Organizations." *Strategic Insights* 10.2.
4. Raymond H. Anderson (Apr. 1978). "Wadi Haddad, Palestinian Hijacking Strategist, Dies." *The New York Times*.
5. Gary LaFree and Laura Dugan (2007). "Introducing the Global Terrorism Database." *Terrorism and Political Violence* 19.2, pp. 181–204.
6. Ihsanullah Tipu Mehsud and Declan Walsh (Aug. 2014). "Hard-Line Splinter Group, Galvanized by ISIS, Emerges From Pakistani Taliban." *The New York Times*.
7. Yuliasri Perdani (Dec. 2013). "Splinter Terrorist Cells May Target Elections: BNPT." *The Jakarta Post*.
8. Jessica Maves Braithwaite and Kathleen Gallagher Cunningham (2020). "When Organizations Rebel: Introducing the Foundations of Rebel Group Emergence (FORGE) Dataset." *International Studies Quarterly* 64.1, pp. 183–193; Michael Woldemariam (2018). *Insurgent Fragmentation in the Horn of Africa: Rebellion and Its Discontents*. Cambridge University Press.
9. Meredith Reid Sarkees and Frank Whelon Wayman (2010). *Resort to War*. Cq Press.
10. Kathleen Gallagher Cunningham (2011). "Divide and Conquer or Divide and Concede: How Do States Respond to Internally Divided Separatists?." *American Political Science Review* 105.2, pp. 275–297.
11. K. G. Cunningham, K. M. Bakke, and L. J. M. Seymour (2012). "Shirts Today, Skins Tomorrow: Dual Contests and the Effects of Fragmentation in Self-Determination Disputes." *Journal of Conflict Resolution* 56.1, pp. 67–93, p. 68.
12. Mia Bloom (2005). *Dying To Kill: The Allure of Suicide Terror*. Columbia University Press, p. 1.
13. Cunningham, Bakke, and Seymour, 2012; Bloom, 2005; Donald L. Horowitz (1985). *Ethnic Groups in Conflict*. University of California Press; Andrew H. Kydd and Barbara F. Walter (2006). "The Strategies of Terrorism." *International Security* 31.1, pp. 49–80; Jack L. Snyder (2000). *From Voting to Violence: Democratization and Nationalist Conflict*. Norton; Monica Duffy Toft (2007). "Getting Religion? The Puzzling Case of Islam and Civil War." *International Security* 31.4, pp. 97–131.

14. Others note that different forms of intragroup or movement factionalization can determine the type of violence and not just its severity or onset. Internal divisions and the inability to control and police member behavior has contributed to civilian victimization in Sierra Leone and El Salvador, respectively. Macartan Humphreys and Jeremy M. Weinstein (2006). "Handling and Manhandling Civilians in Civil War." *American Political Science Review* 100.3, p. 429; Elisabeth Jean Wood (2009). "Armed Groups and Sexual Violence: When Is Wartime Rape Rare?." *Politics & Society* 37.1, pp. 131–161.

15. Barak Mendelsohn (2015). *The Al-Qaeda Franchise: The Expansion of Al-Qaeda and Its Consequences*. Oxford University Press.

16. Christopher W. Blair, Erica Chenoweth, Michael C. Horowitz, Evan Perkoski, and Philip BK Potter (2022). "Honor among Thieves: Understanding Rhetorical and Material Cooperation among Violent Nonstate Actors." *International Organization* 76.1, pp. 164–203.

17. Costantino Pischedda (2018). "Wars within Wars: Why Windows of Opportunity and Vulnerability Cause Inter-Rebel Fighting in Internal Conflicts." *International Security* 43.1, pp. 138–176, p. 154.

18. Jacob Zenn (Feb. 2020). "Boko Haram's Conquest for the Caliphate: How Al Qaeda Helped Islamic State Acquire Territory." *Studies in Conflict & Terrorism* 43.2, pp. 89–122.

19. Audrey Kurth Cronin (2009). *How Terrorism Ends: Understanding the Decline and Demise of Terrorist Campaigns*. Princeton University Press; Mia M. Bloom (2004). "Palestinian Suicide Bombing: Public Support, Market Share, and Outbidding." *Political Science Quarterly* 119.1, pp. 61–88; Bloom, 2005; Horowitz, 1985.

20. Bruce Pirnie and Edward O'Connell (2008). *Counterinsurgency in Iraq (2003–2006)*. Vol. 2. Rand Corporation, p. xiv.

21. Desiree Nilsson (2010). "Turning Weakness into Strength: Military Capabilities, Multiple Rebel Groups and Negotiated Settlements." *Conflict Management and Peace Science* 27.3, pp. 253–271.

22. J. Driscoll (Apr. 2012). "Commitment Problems or Bidding Wars? Rebel Fragmentation as Peace Building." *Journal of Conflict Resolution* 56.1, pp. 118–149.

23. Cunningham, 2011, p. 95.

24. Frances Z. Brown (Oct. 2018). *Dilemmas of Stabilization Assistance: The Case of Syria* Carnegie Endowment for International Peace.

25. Stephen John Stedman (1997). "Spoiler Problems in Peace Settlements." *International Security* 22.2, pp. 5–53, p. 5.

26. David E. Cunningham (2006). "Veto Players and Civil War Duration." *American Journal of Political Science* 50.4, pp. 875–892, p. 895.

27. Ethan Bueno de Mesquita (2005a). "Conciliation, Counterterrorism, and Patterns of Terrorist Violence." *International Organization* 59.01. Ethan Bueno de Mesquita (Apr. 2005b). "The Terrorist Endgame: A Model with Moral Hazard and Learning." *Journal of Conflict Resolution* 49.2, pp. 237–258; Stephen John Stedman (1997). "Spoiler Problems in Peace Settlements." *International Security* 22.2, pp. 5–53; Kelly M. Greenhill and Solomon Major (Jan. 2007). "The Perils of Profiling:

Civil War Spoilers and the Collapse of Intrastate Peace Accords." *International Security* 31.3, pp. 7–40; Wendy Pearlman (2009). "Spoiling Inside and Out: Internal Political Contestation and the Middle East Peace Process." *International Security* 33.3, pp. 79–109.

28. Cunningham, Bakke, and Seymour (2012).

29. This typology focuses on how new armed organizations emerge independently or from other *armed* groups. One could expand this typology almost infinitely to include armed groups emerging (cooperatively or contentiously) from myriad types of organizations. For instance, groups emerging from political parties will likely have some advantages over those emerging from no institutional context whatsoever. Past nonviolent experiences could endow a new armed group with knowledge about organizing, recruiting, and fundraising that can be critically important in the group's formative years. And, as Staniland (2014) argues, groups embedded in local institutions and in society more broadly gain an informational advantage: they can use their relationships to identify and recruit the most committed members, while avoiding those whose commitment is suspect. For this very reason, researchers are currently exploring how these different backgrounds shape armed groups moving forward (e.g., Braithwaite and Cunningham, 2020).

30. Eteri Tsintsadze-Maass and Richard W. Maass (Oct. 2014). "Groupthink and Terrorist Radicalization." *Terrorism and Political Violence* 26.5, pp. 735–758.

31. James D. Fearon and David D. Laitin (2003). "Ethnicity, Insurgency, and Civil War." *American Political Science Review* 97.1, pp. 75–90.

32. Barak Mendelsohn (2011). "Al-Qaeda's Franchising Strategy." *Survival* 53.3, pp. 29–50.

33. Mendelsohn, 2015, p. 30.

34. Ibid.

35. Eric Schmitt (Jan. 2012). "Intelligence Report Lists Iran and Cyberattacks as Leading Concerns." *The New York Times.*

36. This is a well-known insight from literature on military command and control. E.g. Stephen Biddle (2010). *Military Power: Explaining Victory and Defeat in Modern Battle.* Princeton University Press; Ryan Grauer and Michael C. Horowitz (2012). "What Determines Military Victory? Testing the Modern System." *Security Studies* 21.1, pp. 83–112.

37. Michael Woldemariam (2011). "Why Rebels Collide: Factionalism and Fragmentation in African Insurgencies." PhD thesis. Princeton University, p. 25.

38. Though there is substantially more on the related question of why political parties develop their own parallel militant wings. E.g. Anisseh Van Engeland and Rachael M. Rudolph (Apr. 2016). From *Terrorism to Politics.* Routledge. James A. Piazza (Feb. 2010). "Terrorism and Party Systems in the States of India." *Security Studies* 19.1, pp. 99–123; Leonard Weinberg and William Eubank (1990). "Political Parties and the Formation of Terrorist Groups." *Terrorism and Political Violence* 2.2, pp. 125–144.

39. Assaf Moghadam (Mar. 2003). "Palestinian Suicide Terrorism in the Second Intifada: Motivations and Organizational Aspects." *Studies in Conflict & Terrorism* 26.2, pp. 65–92, p. 82.

40. Moghadam, 2003; Luther P. Gerlach (2001). "The Structure of Social Movements: Environmental Activism and Its Opponents." *Networks and Netwars: The Future of Terror, Crime, and Militancy*, pp. 289–310, p. 82.

41. Few studies engage with the definitional and observational complexity surrounding organizational splits. A notable exception is: Woldemariam, 2011, p. 35.

42. Ethan Bueno de Mesquita (Dec. 2008). "Terrorist Factions." *Quarterly Journal of Political Science* 3.4, pp. 399–418.

43. Jacob Zenn (Feb. 2014). "Leadership Analysis of Boko Haram and Ansaru in Nigeria." *Combating Terrorism Center at West Point*. CTC Sentinel 7.2, p. 26.

44. Zenn, 2020.

45. Erin M. Kearns, Allison E. Betus, and Anthony F. Lemieux (2019). "Why Do Some Terrorist Attacks Receive More Media Attention than Others?" *Justice Quarterly* 36.6, pp. 985–1022.

46. Michael Jetter (2019). "More Bang for the Buck: Media Coverage of Suicide Attacks." *Terrorism and Political Violence* 31.4, pp. 779–799.

47. Lizzie Dearden (Jan. 2017). "Al-Qaeda Leader Denounces Isis' 'Madness and Lies' as Two Terrorist Groups Compete for Dominance." *The Independent*.

48. Rob Crilly (Apr. 2019). "Trump Admits He Can't Say ISIS Defeated Because 'whack Jobs' Could Strike." *Washington Examiner*.

49. Kathy Frankovic (Mar. 2019). *What Is Terrorism? For Americans, Who Matters More than What*. YouGov.

50. Gordon H. McCormick (2003). "Terrorist Decision Making." *Annual Review of Political Science* 6.1, pp. 473–507; Todd Sandler and Daniel G. Arce (2007). "Terrorism: A Game-Theoretic Approach." *Handbook of Defense Economics*. Ed. by Todd Sandler and Keith Hartley. Vol. 2. *Handbook of Defense Economics Defense in a Globalized World*. Elsevier, pp. 775–813; Martha Crenshaw (1981). "The Causes of Terrorism." *Comparative Politics* 13.4, pp. 379–399.

51. There is considerable debate regarding suicide terrorism, however, with some scholars arguing that this behavior classifies as irrational. Those who believe that suicide bombers are rational (or at least not irrational) commonly cite three facts: first, that viewing the act as a self-sacrifice for a greater good makes it easier to logically justify the act; second, that a bomber's family will often receive compensation, which is more than what an individual could provide on his or her own; and third, that group pressure and socialization might be the key link between a rational individual and a suicide bomber, though again this would not make them irrational. E.g., Adam Lankford (2010). "Do Suicide Terrorists Exhibit Clinically Suicidal Risk Factors? A Review of Initial Evidence and Call for Future Research." *Aggression and Violent Behavior* 15.5, pp. 334–340; Adam Lankford (2011). "Could Suicide Terrorists Actually Be Suicidal?" *Studies in Conflict & Terrorism* 34.4, pp. 337–366; Ariel Merari (1990). "The Readiness to Kill and Die: Suicidal Terrorism in the Middle East." *Origins of terrorism* 192; Ariel Merari (2004). "Suicide Terrorism." *Assessment, Treatment, and Prevention of Suicidal Behavior*, pp. 431–454; Mohammed M. Hafez (2006). "Rationality, Culture, and Structure in the Making of Suicide Bombers: A Preliminary Theoretical Synthesis and Illustrative Case Study." *Studies in Conflict &*

Terrorism 29.2, pp. 165–185; Bloom, 2005; Jerrold M. Post (2005). "When Hatred Is Bred in the Bone: Psycho-Cultural Foundations of Contemporary Terrorism." *Political Psychology* 26.4, pp. 615–636; Robert A. Pape (2003). "The Strategic Logic of Suicide Terrorism." *American Political Science Review* 97.3, pp. 343–361; Robert A. Pape (2005). *Dying to Win: The Strategic Logic of Terrorism*. Random House; Louise Richardson (2006). *What Terrorists Want: Understanding the Enemy, Containing the Threat*. Random House Digital, Inc.

52. Gaetano Joe Ilardi (2009). "The 9/11 Attacks—A Study of Al Qaeda's Use of Intelligence and Counter Intelligence." *Studies in Conflict & Terrorism* 32.3, 171–187.

53. Walter Enders and Todd Sandler (1993). "The Effectiveness of Antiterrorism Policies: A Vector-Autoregression-Intervention." *American Political Science Review* 87.04, pp. 829–844.

54. Bloom, 2004.

55. It is important to caveat that rationality does not imply that armed groups will always make *correct* decisions, and they may not achieve their desired political goals. Rationality implies that groups make logical decisions accounting for their biases and sometimes limited information. Even a rational armed group could conduct a spectacular failure of an attack or even be defeated within months of forming. Researchers generally apply this same conception of rationality to states. E.g. James D. Fearon (1995). "Rationalist Explanations for War." *International Organization* 49.03, pp. 379–414; Kydd and Walter, 2006.

56. Seth G. Jones and Martin C. Libicki (2008). *How Terrorist Groups End: Lessons for Countering Al Qa'ida*. Rand Corporation.

57. John F. Morrison (2013). *Origins and Rise of Dissident Irish Republicanism: The Role and Impact of Organizational Splits*. A&C Black, p. 13.

58. Hart and Van Vugt, 2006, p. 392.

59. Bruno Dyck (1997). "Exploring Organizational Family Trees: A Multigenerational Approach for Studying Organizational Births." *Journal of Management Inquiry* 6.3, pp. 222–233.

60. David Besanko et al. (2009). *Economics of Strategy*. John Wiley & Sons.

61. Steve Lohr (Dec. 2018). "G.E. to Spin Off Its Digital Business." *The New York Times*.

62. Raphael Zariski (1960). "Party Factions and Comparative Politics: Some Preliminary Observations." *Midwest Journal of Political Science*, pp. 27–51, p. 33.

63. Rodney Stark and William Sims Bainbridge (1985). *The Future of Religion: Secularization, Revival, and Cult Formation*. University of California Press, p. 102.

64. For an exemplary exception, see: Torun Dewan and Francesco Squintani (2012). "The Role of Party Factions: An Information Aggregation Approach." Manuscript available online at http://personal. lse. ac. uk/DEWANta.

65. Patrick Köllner (2004). "Factionalism in Japanese Political Parties Revisited or How Do Factions in the LDP and the DPJ Differ?" *Japan Forum* 16.1, pp. 87–109, pp. 89–90.

66. Braithwaite and Cunningham, 2020.

67. Woldemariam, 2018.

68. Nicholai Hart Lidow (2016). *Violent Order: Understanding Rebel Governance through Liberia's Civil War*. Cambridge University Press.

69. For an excellent overview of this literature, see Kent Eriksson and Sylvie Chetty (2003). "The Effect of Experience and Absorptive Capacity on Foreign Market Knowledge." *International Business Review* 12.6, pp. 673–695.

70. For seminal work on this topic, see: George P. Huber (1991). "Organizational Learning: The Contributing Processes and the Literatures." *Organization Science* 2.1, pp. 88–115.

71. Johan Bruneel, Helena Yli-Renko, and Bart Clarysse (2010). "Learning from Experience and Learning from Others: How Congenital and Interorganizational Learning Substitute for Experiential Learning in Young Firm Internationalization." *Strategic Entrepreneurship Journal* 4.2, p. 164, p. 167.

72. H. J. Sapienza et al. (Oct. 2006). "A Capabilities Perspective on the Effects of Early Internationalization on Firm Survival and Growth." *Academy of Management Review* 31.4, pp. 914–933; Jan Johanson and Jan-Erik Vahlne (1990). "The Mechanism of Internationalisation." *International Marketing Review* 7.4.

73. Sapienza et al., 2006; Helena Yli-Renko, Erkko Autio, and Harry J. Sapienza (2001). "Social Capital, Knowledge Acquisition, and Knowledge Exploitation in Young Technology-Based Firms." *Strategic Management Journal* 22.6–7, pp. 587–613.

74. Sapienza et al., 2006.

75. Ranjay Gulati and Maxim Sytch (2008). "Does Familiarity Breed Trust? Revisiting the Antecedents of Trust." *Managerial and Decision Economics* 29.2–3, p. 165; Indre Maurer (2010). "How to Build Trust in Inter-Organizational Projects: The Impact of Project Staffing and Project Rewards on the Formation of Trust, Knowledge Acquisition and Product Innovation." *International Journal of Project Management* 28.7, pp. 629–637.

76. Stephen Bloomer (1993). "The History and Politics of the I.R.S.P & I.N.L.A.: From 1974 to the Present Day." MSc thesis, Queens University Belfast, p. 6.

77. George W. Bush (Dec. 2005). "Statement of the President George W. Bush on the Importance of the USA Patriot In Combatting Terrorism." The White House.

78. Barack H. Obama (Oct. 2013). "President Obama Speaks at the Installation of FBI Director James Comey." The White House.

79. David H. Petraeus and James F. Amos (2009). *US Army US Marine Corps Counterinsurgency Field Manual*. Signalman Publishing.

80. Ibid.

81. Ibid.

82. Biddle, 2010.

83. Shapiro, 2013.

84. Stephen Peter Rosen (1988). "New Ways of War: Understanding Military Innovation." *International Security* 13.1, pp. 134–168.

85. Shapiro, 2013.

86. Petraeus and Amos, 2009.

87. So how has fragmentation become a dominant counterterrorism and counterinsurgency strategy if it is actually so uncertain? The success of divide and conquer approaches in interstate wars has biased the expectation of its utility in subnational

conflicts. While there is a wealth of information attesting to its efficacy against states, there is much less evidence suggesting it works against less developed and less bureaucratic enemies who do not rely as heavily on control and coordination. One explanation for its adoption is the influence of cognitive bias, which basically means that individuals' preexisting beliefs shape how they process information. this specific case, when military commanders believe that fragmentation strategies are effective, because of their knowledge of interstate wars, they are already more likely to utilize it in subnational conflicts regardless of whatever future information might arrive. Lyall and Wilson find that related inter-state war biases have increased the rate at which states lose wars to insurgent organizations in the twentieth century. Also see: Dale Griffin and Amos Tversky (1992). "The Weighing of Evidence and the Determinants of Confidence." *Cognitive Psychology* 24.3, pp. 411–435. Judith Orasanu, Roberta Calderwood, and Caroline E. Zsambok, eds (1993). "Decision making in action: Models and methods." Ablex Publishing. Charles R. Schwenk (1984). "Cognitive Simplification Processes in Strategic Decision-Making." *Strategic Management Journal* 5.2, pp. 111–128; Jason Lyall and Isaiah Wilson (2009). "Rage against the Machines: Explaining Outcomes in Counterinsurgency Wars." *International Organization* 63.1, pp. 67–106.

88. Ekkart Zimmerman (1980). "Macro-Comparative Research on Political Protest." *Handbook of Political Conflict: Theory and Research*, pp. 167–237, p. 191.
89. George Simmel (1955). "Conflict (KH Wolff, Trans.)" The Free Press.
90. Lewis A. Coser (1956). *The Functions of Social Conflict.* The Free Press.
91. Mancur Olson (1982). *The Rise and Fall of Nations: Economic Growth, Stagflation and Social Rigidities.* Yale University Press.
92. Charles Tilly (1978). *From Mobilization to Revolution.* McGraw-Hill.
93. Jacob N. Shapiro (2013). *The Terrorist's Dilemma: Managing Violent Covert Organizations.* Princeton University Press.
94. V. Asal, M. Brown, and A. Dalton (2012). "Why Split? Organizational Splits among Ethnopolitical Organizations in the Middle East." *Journal of Conflict Resolution* 56.1, pp. 94–117.
95. Theodore McLauchlin and Wendy Pearlman (2012). "Out-Group Conflict, In-Group Unity? Exploring the Effect of Repression on Intramovement Cooperation." *Journal of Conflict Resolution* 56.1, pp. 41–66, p. 44.
96. Woldemariam, 2011.
97. Christia finds a similar pattern with regards to intergroup alliances during multiparty civil wars. While not specifically focused on organizational unity, the close cooperation between armed groups does present a similar scenario that we can potentially learn from. Her research reveals that alliances are a function of power relations and post-conflict payouts. Applying this to intraorganizational dynamics, it suggests that selfish motivations and potential pay-outs could motivate splinters to break away. Fotini Christia (2012). *Alliance Formation in Civil Wars.* Cambridge University Press, pp. 239–240.
98. Woldemariam, 2011, pp. 3–4.

99. Paul D. Kenny (2010). "Structural Integrity and Cohesion in Insurgent Organizations: Evidence from Protracted Conflicts in Ireland and Burma." *International Studies Review* 12.4, pp. 533–555.

100. Kenny, 2010, p. 552.

101. Eric S. Mosinger (2018). "Brothers or Others in Arms? Civilian Constituencies and Rebel Fragmentation in Civil War." *Journal of Peace Research* 55.1, pp. 62–77; Austin C. Doctor (2020). "A Motion of No Confidence: Leadership and Rebel Fragmentation." *Journal of Global Security Studies* 5.4, pp. 598–616; Pischedda, 2018.

102. Paul Staniland (2010). "Explaining Cohesion, Fragmentation, and Control in Insurgent Groups." PhD thesis. Massachusetts Institute of Technology, p. 12.

103. Asal, Brown, and Dalton, 2012.

104. Shapiro, 2013; Lee J. M. Seymour, Kristin M. Bakke, and Kathleen Gallagher Cunningham (2016). "E Pluribus Unum, Ex Uno Plures: Competition, Violence, and Fragmentation in Ethnopolitical Movements." *Journal of Peace Research* 53.1, pp. 3–18.

105. Although they may create "or exacerbating internal conflict." Deborah B. Balser (1997). "The Impact of Environmental Factors on Factionalism and Schism in Social Movement Organizations." *Social Forces* 76.1, pp. 199–228, p. 201.

106. Finke and Scheitle (2009). "Understanding Schisms: Theoretical Explanations for Their Origins". In James R. Lewis and Sarah M. Lewis (Eds.), *Sacred Schisms: How Religions Divide*, pp. 11–34. P. 12.

107. Kydd and Walter, 2006.

108. Rohan Gunaratna and Khuram Iqbal (2012). *Pakistan: Terrorism Ground Zero.* Reaktion Books.

109. Ibid.

110. Aarish Ulla Khan (2005). *The Terrorist Threat and the Policy Response in Pakistan.* SIPRI Policy Paper No. 11.

111. *Peshawar School Massacre Splits Afghan, Pakistani Militant* Groups" (December 19, 2014). NBC News. https:// www.nbcnews.com/storyline/pakistan-school-massacre/peshawar-school-massacre-splits-afghan-pakistani-militant-groups

112. Though, to be sure, there is significant variation even within civil wars, with acts of terrorism not uncommon. E.g., Jessica A. Stanton (2013). "Terrorism in the Context of Civil War." *The Journal of Politics* 75.4, pp. 1009–1022.

113. For a few notable examples, see Michael C. Horowitz (2010). "Nonstate Actors and the Diffusion of Innovations: The Case of Suicide Terrorism." *International Organization* 64.1; Shapiro, 2013; Schweitzer, 2011; Brian J. Phillips (2013). "Terrorist Group Cooperation and Longevity." *International Studies Quarterly*; Brian J. Phillips (2015). "Enemies with Benefits? Violent Rivalry and Terrorist Group Longevity." *Journal of Peace Research* 52.1, pp. 62–75; Staniland, 2010; David B. Carter (2012). "A Blessing or a Curse? State Support for Terrorist Groups." *International Organization* 66.01, pp. 129–151; Joseph K. Young and Laura Dugan (Apr. 2014). "Survival of the Fittest: Why Terrorist Groups Endure." *Perspectives on Terrorism* 8.2.

114. Though there are certainly many degrees between the two. For example, within the network category experts often cite three different models of group structure: the

chain, hub-and-spoke, and all-channel network. These are thoroughly discussed in: Brian A. Jackson (2006). "Groups, Networks, or Movements: A Command-and-Control-Driven Approach to Classifying Terrorist Organizations and Is Application to Al Qaeda." *Studies in Conflict & Terrorism* 29.3, pp. 241–262.

115. Shapiro, 2013.
116. Shapiro, 2013; Jenna Jordan (2009). "When Heads Roll: Assessing the Effectiveness of Leadership Decapitation." *Security Studies* 18.4, pp. 719–755; Bryan C. Price (2012). "Targeting Top Terrorists: How Leadership Decapitation Contributes to Counterterrorism." *International Security* 36.4, pp. 9–46; Patrick B. Johnston (2012). "Does Decapitation Work? Assessing the Effectiveness of Leadership Targeting in Counterinsurgency Campaigns." *International Security* 36.4, pp. 47–79; Jenna Jordan (2014). "Attacking the Leader, Missing the Mark." *International Security* 38.4, pp. 7–38.
117. Joshua Kilberg (2011). "Organizing for Destruction: Does Organizational Structure Affect Terrorist Group Behavior and Success?" PhD dissertation, Norman Parerson School of International Affairs, Carleton University, Canada.
118. Horowitz, 2010.
119. Everett M. Rogers (2010). *Diffusion of Innovations.* Simon and Schuster.

Chapter 2

1. Brian J. Phillips (2019). "Do 90 Percent of Terrorist Groups Last Less than a Year? Updating the Conventional Wisdom." *Terrorism and Political Violence* 31.6, pp. 1255–1265.
2. Louis R. Pondy (1967). "Organizational Conflict: Concepts and Models." *Administrative Science Quarterly*, pp. 296–320; Stephen Worchel et al. (1992). *Group Process and Productivity.* SAGE Publications; Stephen Worchel (1998). *Social Identity: International Perspectives.* Sage; Kelsy Kretschmer (2013). "Factions/Factionalism." *The Wiley-Blackwell Encyclopedia of Social and Political Movements.*
3. Weinberg and Eubank, 1990.
4. Ibid, p. 356.
5. Zariski, 1960, p. 33.
6. Much research has leveraged this analogy to better understand armed groups in particular. E.g. Weinberg and Eubank, 1990; Anthony Richards (2001). "Terrorist Groups and Political Fronts: The IRA, Sinn Fein, the Peace Process and Democracy." *Terrorism and Political Violence* 13.4, pp. 72–89; Leonard Weinberg (1991). "Turning to Terror: The Conditions under Which Political Parties Turn to Terrorist Activities." *Comparative Politics* 23.4, pp. 423–438; Leonard Weinberg, Ami Pedahzur, and Arie Perliger (2008). *Political Parties and Terrorist Groups.* Vol. 10. Routledge; Erica Chenoweth (2010). "Democratic Competition and Terrorist Activity." *Journal of Politics* 72.1, pp. 16–30.
7. Zariski, 1960, p. 19.

8. In other words, there is no clearly defined leadership, leadership structures, or succession plans. Dennis C. Beller and Frank P. Belloni (1978). "Party and Faction: Modes of Political Competition." *Faction Politics: Political Parties and Factionalism in Comparative Perspective*, p. 417, pp. 419–422.

9. Like Sendero Luminoso that emerged from the Communist Party of Peru (Weinberhg, 1991).

10. Aymenn Al-Tamimi (2017). "The Formation of Hay'at Tahrir Al-Sham and Wider Tensions in the Syrian Insurgency." *CTC Sentinel* 10.2; *The Armed Opposition in Northwest Syria* (May 2020). The Carter Center.

11. *Al-Nusra Front (Hayat Tahrir Al-Sham)* (n.d.). Counter Extremism Project.

12. While this finding is supported by research on group psychology and organization dynamics, it also finds support in existing research on when armed groups split. E.g. Victor Asal and R. Karl Rethemeyer (2008). "The Nature of the Beast: Organizational Structures and the Lethality of Terrorist Attacks." *Journal of Politics* 70.2, 437–449.

13. Morrison, 2013.

14. Michael A. Hogg and Joseph A. Wagoner (2017). "Uncertainty–Identity Theory." *The International Encyclopedia of Intercultural Communication*, pp. 1–9; Matthew J. Hornsey (2008). "Social Identity Theory and Self-Categorization Theory: A Historical Review." *Social and Personality Psychology Compass* 2.1, pp. 204–222; Saul McLeod (2008). "Social Identity Theory." *Simply Psychology*; John C. Turner et al. (1987). *Rediscovering the Social Group: A Self-Categorization Theory*. Basil Blackwell.

15. Clifford R. Mynatt, Michael E. Doherty, and Ryan D. Tweney (1977). "Confirmation Bias in a Simulated Research Environment: An Experimental Study of Scientific Inference." *Quarterly Journal of Experimental Psychology* 29.1, pp. 85–95; Joshua Klayman (1995). "Varieties of Confirmation Bias." *Psychology of Learning and Motivation* 32, pp. 385–418.

16. Of course, research also finds that subgroups can form around other connections not linked to individuals' beliefs. These connections can include kinship, shared ethnicity, or birthplace. Subgroups formed around these connections would not count as factions here since they are not bound together by a shared purpose and do not work together to achieve that goal. Rather, they are simply bound by a shared characteristic that might provide familiarity and reassurance. Furthermore, it is unclear if kinship-based factions will have the same effects on organizational cohesion as belief-based factions. In some cases, as Parkinson shows, these subgroups can actually serve valuable purposes. E.g., Sarah Elizabeth Parkinson (2013). "Organizing Rebellion: Rethinking High-Risk Mobilization and Social Networks in War." *American Political Science Review* 107.3, pp. 418–432.

17. John C. Turner and Katherine J. Reynolds (2011). "Self-Categorization Theory." *Handbook of Theories in Social Psychology* 2.1, pp. 399–417.

18. H. Turner Tajfel and John Turner (1986). "The Social Identity Theory of Intergroup Behavior." *Psychology of Intergroup Relations*, pp. 7–24; Turner et al., 1987.

19. Balser, 1997; Kretschmer, 2013.

20. Thomas E. Shriver and Chris Messer (2009). "Ideological Cleavages and Schism in the Czech Environmental Movement." *Human Ecology Review*, pp. 161–171, p. 170.

21. Steven J. Hood (1996). "Political Change in Taiwan: The Rise of Kuomintang Factions." *Asian Survey* 36.5, pp. 468–482.

22. Clem Lloyd and Wayne Swan (1987). "National Factions and the ALP." *Politics* 22.1, pp. 100–110.

23. Charles Lister (Aug. 2019). "What's next for Al-Qaeda after the Death of Hamza Bin Laden." *Middle East Institute*.

24. Bruno Dyck and Frederick A. Starke (1999). "The Formation of Breakaway Organizations: Observations and a Process Model." *Administrative Science Quarterly* 44.4, pp. 792–822, p. 807.

25. Andrew M. Carton (2011). "A Theory, Measure, and Empirical Test of Subgroups in Work Teams." PhD thesis. Duke University, p. 11.

26. Zariski, 1960, p. 33.

27. Ahmed S. Hashim (2014). "The Islamic State: From Al-Qaeda Affiliate to Caliphate." *Middle East Policy* 21.4, pp. 69–83.

28. William McCants (2015). *The ISIS Apocalypse: The History, Strategy, and Doomsday Vision of the Islamic State.* Macmillan.

29. Unfortunately, evidence of the impending split still did not help the US and coalition allies prevent or really prepare for this new threat. See: Nelly Lahoud and Liam Collins (2016). "How the CT Community Failed to Anticipate the Islamic State." *Democracy and Security* 12.3, pp. 199–210.

30. Shapiro, 2013, p. 26.

31. Mayer N. Zald and John D. McCarthy (1979). "Social Movement Industries: Competition and Cooperation among Movement Organizations." CRSO Working Paper No. 201, Center for Research on Social Organization, University of Michigan.

32. Mayer N. Zald and Roberta Ash (1966). "Social Movement Organizations: Growth, Decay and Change." *Social Forces* 44.3, pp. 327–341.

33. William A. Gamson (1975). *The Strategy of Social Protest.* Dorsey Press.

34. Morrison, 2013, p. 19.

35. Hart and Van Vugt, 2006.

36. Rebecca H. Best and Navin A. Bapat (Jan. 2018). "Bargaining with Insurgencies in the Shadow of Infighting." *Journal of Global Security Studies* 3.1, pp. 23–37.

37. Morrison, 2013; Woldemariam, 2011; Doctor, 2020.

38. Morrison, 2013.

39. Young and Dugan, 2014.

40. Jack Holland and Henry McDonald (1994). *INLA, Deadly Divisions: The Story of One of Ireland's Most Ruthless Terrorist Organisations.* Torc, p. 33.

41. Lawahez Jabari, Saphora Smith, and Yuliya Talmazan (November 13, 2018). "Gaza Militants Announce Ceasefire with Israel in Bid to Avert War." *NBC News*.

42. Mariam Safi (Dec. 2007). *Talking to "Moderate" Taliban.* Institute of Peace and Conflict Studies.

43. Ben Brandt (July 2010). *The Punjabi Taliban*. Combating Terrorism Center at West Point. Chap. Sentinel Article.
44. Dyck and Starke, 1999; Kretschmer, 2013, p. 89.
45. Montgomery Sapone (2000). "Ceasefire: The Impact of Republican Political Culture on the Ceasefire Process in Northern Ireland." *Peace and Conflict Studies* 7.1, 24–51.
46. Greg Miller (May 2011). "Al-Qaeda Confirms Osama Bin Laden's Death, Vows Retaliation." *Washington Post*.
47. Morrison, 2013.
48. Morrison, 2013; Jones and Libicki, 2008.
49. Tulia G. Falleti and James Mahoney (2015). "The Comparative Sequential Method." *Advances in Comparative Historical Analysis: Resilience, Diversity, and Change*, pp. 211–239, p. 222.
50. Dora C. Lau and J. Keith Murnighan (1998). "Demographic Diversity and Faultlines: The Compositional Dynamics of Organizational Groups." *Academy of Management Review* 23.2, pp. 325–340; Hart and Van Vugt, 2006.
51. Hart and Van Vugt, 2006.
52. Bueno de Mesquita, 2005a; Morrison, 2013; Robert P. Clark (1990). *Negotiating with ETA: Obstacles to Peace in the Basque Country, 1975–1988*. University of Nevada Press; Robert P. Clark (1984). *The Basque Insurgents: ETA, 1952–1980*. University of Wisconsin Press Madison; Andrew G. Reiter (2015). "Does Spoiling Work? Assessing the Impact of Spoilers on Civil War Peace Agreements." *Civil Wars* 17.1, pp. 89–111.
53. Bueno de Mesquita, 2008, p. 403.
54. Rex A. Hudson (Sept. 1999). *The Sociology and Psychology of Terrorism: Who Becomes a Terrorist and Why?* Federal Research Division, Library of Congress.
55. Paul Staniland (2012). "Between a Rock and a Hard Place: Insurgent Fratricide, Ethnic Defection, and the Rise of Pro-State Paramilitaries." *Journal of Conflict Resolution* 56.1, pp. 16–40.
56. Daveed Gartenstein-Ross (Sept. 2009). "The Strategic Challenge of Somalia's Al-Shabaab." *Middle East Quarterly*; Rob Wise (2011). *Al Shabaab*. Center for Strategic and International Studies.
57. Armin Rosen (Mar. 2012). "The Warlord and the Basketball Star: A Story of Congo's Corrupt Gold Trade." *The Atlantic*.
58. Brandt, 2010.
59. Kathleen Gallagher Cunningham (2016). *Understanding Fragmentation in Conflict and Its Impact on Prospects for Peace*. Centre for Humanitarian Dialogue: Oslo Forum.
60. Relatedly, research finds that personalist regimes are some of the weakest among all authoritarian structures. E.g., Abel Escribà-Folch and Joseph Wright (2010). "Dealing with Tyranny: International Sanctions and the Survival of Authoritarian Rulers." *International Studies Quarterly* 54.2, pp. 335–359; Barbara Geddes (1999). "Authoritarian Breakdown: Empirical Test of a Game Theoretic Argument." *Annual Meeting of the American Political Science Association*. Atlanta, Georgia.

61. Alan Greenblatt (May 2011). "Without Bin Laden, How Dangerous Is Al-Qaida?." *NPR*.

62. Martha Crenshaw (2007). "Explaining Suicide Terrorism: A Review Essay." *Security Studies* 16.1, pp. 133–162.

63. Bruce Hoffman and Gordon H. Mccormick (July 2004). "Terrorism, Signaling, and Suicide Attack." *Studies in Conflict & Terrorism* 27.4, pp. 243–281, p. 259.

64. Martha Crenshaw (1990). "Questions to Be Answered, Research to Be Done, Knowledge to Be Applied." *Origins of Terrorism: Psychologies, Ideologies, Theologies, States of Mind*. Ed. by W. Reich. Cambridge University Press, pp. 247–260, p. 250.

65. Hudson, 1999.

66. Bueno de Mesquita, 2005a.

67. Hudson, 1999.

68. Further demonstrating the variation that exists among the members of an armed group is their reasons for joining. Some are motivated by greed, others by respect, others by a particular cause such as nationalism or ideological fervor, and some join simply because their friends are doing it, too. This motivational variation is well studied with regard to members of the Islamic State. E.g., Macartan Humphreys and Jeremy M. Weinstein (2008). "Who Fights? The Determinants of Participation in Civil War." *American Journal of Political Science* 52.2, pp. 436–455; Vera Mironova (2019). "Who Are the ISIS People?." *Perspectives on Terrorism* 13.1, pp. 32–39.

69. Morrison, 2013, pp. 14–15.

70. Matthew Levitt (2009). "Hamas's Ideological Crisis." *Current Trends in Islamist Ideology* 9, pp. 80–95, p. 93.

71. Madeline Townsend (October 20, 2016). "After the Violence Abated: The Aftermath of Sendero Luminoso." Panoramas, University of Pittsburgh.

72. Anders Strindberg (2000). "The Damascus-Based Alliance of Palestinian Forces: A Primer." *Journal of Palestine Studies* 29.3, pp. 60–76, p. 66.

73. This would be a strategic dispute, and not ideological, since it concerned the organization's ultimate goals. There was no disagreement over the interpretation of Islamic doctrine or thought. See, for instance: Thomas G. Wilson Jr. (2009). *Extending the Autonomous Region in Muslim Mindanao to the Moro Islamic Liberation Front: A Catalyst for Peace*. Fort Leavenworth, Kansas: School of Advanced Military Studies United States Army Command and General Staff College.

74. Glenn R. Carroll (1984). "Organizational Ecology." *Annual Review of Sociology* 10.1, pp. 71–93; Joel A. C. Baum and Jitendra V. Singh (1994). "Organizational Niches and the Dynamics of Organizational Founding." *Organization Science* 5.4, pp. 483–501; Jitendra V. Singh (1994). "Organizational Niches and the Dynamics of Organizational Mortality." *American Journal of Sociology* 100.2, pp. 346–380; John Freeman and Michael T. Hannan (1983). "Niche Width and the Dynamics of Organizational Populations." *American Journal of Sociology* 88.6, pp. 1116–1145; Michael T. Hannan, Glenn R. Carroll, and László Pólos (2003). "The Organizational Niche." *Sociological Theory* 21.4, pp. 309–340; Michael T. Hannan and John Freeman (1977). "The Population Ecology of Organizations." *American Journal of Sociology* 82.5, pp. 929–964.

75. Pamela A. Popielarz and Zachary P. Neal (2007). "The Niche as a Theoretical Tool." *Annual Review of Sociology* 33, p. 65.

76. Baum and Singh, 1994.

77. Stanislav D. Dobrev, Tai-Young Kim, and Michael T. Hannan (2001). "Dynamics of Niche Width and Resource Partitioning." *American Journal of Sociology* 106.5, pp. 1299–1337.

78. Glenn R. Carroll and Anand Swaminathan (Nov. 2000). "Why the Microbrewery Movement? Organizational Dynamics of Resource Partitioning in the U.S. Brewing Industry." *American Journal of Sociology* 106.3, pp. 715–762.

79. Brandice Canes-Wrone, David W. Brady, and John F. Cogan (2002). "Out of Step, Out of Office: Electoral Accountability and House Members' Voting." *American Political Science Review* 96.1, pp. 127–140.

80. Hannan, Carroll, and Pólos, 2003; Hannan and Freeman, 1977; Christopher P. Scheitle (2007). "Organizational Niches and Religious Markets: Uniting Two Literatures." *Interdisciplinary Journal of Research on Religion* 3, pp. 1–29.

81. Baum and Singh, 1994.

82. Yen Hsu. "Design innovation and marketing strategy in successful product competition." *Journal of Business & Industrial Marketing* 26.4, pp. 223–236.

83. Olav Sorenson, Susan McEvily, Charlotte Rongrong Ren, and Raja Roy (2006). "Niche width revisited: Organizational scope, behavior and performance." *Strategic Management Journal* 27. 10, pp. 915–936.

84. Dobrev, Kim, and Hannan, 2001.

85. Scheitle, 2007.

86. Haris Aziz, Felix Brandt, and Paul Harrenstein (2013). "Pareto Optimality in Coalition Formation." *Games and Economic Behavior* 82, pp. 562–581.

87. Allard Duursma and Feike Fliervoet (2020). "Fueling Factionalism? The Impact of Peace Processes on Rebel Group Fragmentation in Civil Wars." *Journal of Conflict Resolution* 65.4, pp. 788–812.

88. Donatella Della Porta (2006). *Social Movements, Political Violence, and the State: A Comparative Analysis of Italy and Germany.* Cambridge University Press, p. 109.

89. Jeremy M. Weinstein (2005). "Resources and the Information Problem in Rebel Recruitment." *Journal of Conflict Resolution* 49.4, pp. 598–624; Humphreys and Weinstein, 2008.

90. The size of an organization's niche—in whichever realm it operates—has important consequences for the group's inner workings. Among religious groups, for instance, "Having a small niche allows religious organizations to effectively market their goods to a specific segment of the population...and allows individuals to create strong ties within the organization" (Scheitle, 2007, p. 22).

91. Frances J. Milliken and Luis L. Martins (1996). "Searching for Common Threads: Understanding the Multiple Effects of Diversity in Organizational Groups." In *Academy of Management Review* 21.2, pp. 402–433.

92. Scheitle, 2007, p. 5.

93. Gates, 2002, p. 115.

94. Shapiro, 2013.

95. Martha Crenshaw (2000). "The Psychology of Terrorism: An Agenda for the 21st Century." *Political Psychology* 21.2, pp. 405–420; Scott Gates (2002). "Recruitment and Allegiance: The Microfoundations of Rebellion." *Journal of Conflict Resolution* 46.1, pp. 111–130.

96. Shapiro, 2013.

97. Walter Enders and Xuejuan Su (2007). "Rational Terrorists and Optimal Network Structure." *Journal of Conflict Resolution* 51.1, pp. 33–57, p. 38.

98. Steven Hutchison and Pat O'Malley (2007). "How Terrorist Groups Decline." *Trends in Terrorism Series*. ITAC, p. 3.

99. Freeman and Hannan, 1983, p. 1143.

100. Christia, 2012.

101. Shapiro, 2013, p. 26.

102. Clark, 1984, p. 35.

103. Ibid., pp. 48–49.

104. This turnover, in its own way, exacerbated divisions within the organization. With the old guard going into hiding, exile, or in prison, new leaders sought to exert control and achieve their own particular visions for the group. While ETA had seemingly settled its political ideology at the 5th Assembly in 1968, the new leadership reopened many of these debates (Clark, 1984; Clark, 1990).

105. Jose L. Rodriguez Jimenez and Luis José (2009). "Los terrorismos en la crisis del Franquismo y en la transición política a la democracia." *Historia del presente* 13, pp. 133–151.

106. Manuel Cerdán (Oct. 2013). *Matar a Carrero: la conspiración: Toda la verdad sobre el asesinato del delfín de Franco*. Penguin Random House Grupo Editorial España; Winter, Ulrich (2016). "Como después de una detonación cambia el silencio...: el atentado de ETA contra Carrero Blanco, la acción directa y la cronopolítica de la resistencia antifranquista," in *Atentado contra Carrero Blanco como lugar de (no-)memoria : narraciones históricas y representaciones culturales* Iberoamericana Editorial Vervuert, pp. 125–138.

107. Clark, 1990, p. 78.

108. Ibid., pp. 79–80.

109. Author's translation. Gaizka Fernández Soldevilla (2013). *Héroes, Heterodoxos y Traidores*. Tecnos, p. 72.

110. Gaizka Fernández Soldevilla (2016). *La Voluntad Del Gudari: Génesis y Metástasis de La Violencia de ETA*. Tecnos, p. 193.

111. Antonio Elorza and José Mari Garmendia (2006). *La Historia de ETA*. Temas de Hoy, p. 256.

112. Author's translation. The original is as follows: Hace dos anos...que ETA se vió dividida con gran dolor y confusión por parte del Pueblo Vasco. Las razones de la excisión fueron las diferencias respeto a la forma de estructurar la organización, el diferente modo de ver la relación entre las cauces organizativos de que se dotaban las diferentes manifestaciones que toma la lucha popular.... Las constantes cáidas surfridas por la organización político-militar mostraron palpablemente la imposibiliad de ejercer dirección política, por parte de una organización que practica a

la vez la lucha armada y la acción de masas en la etapa actual de la revolución. *Documents Y* (1981). Vol. 1. Hordago, p. 21805.

113. Ibid., Vol. 1, p. 20.

114. Clark, 1990, p. 77.

115. Charles W. Mahoney (Feb. 2020). "Splinters and Schisms: Rebel Group Fragmentation and the Durability of Insurgencies." *Terrorism and Political Violence* 32.2, pp. 345–364.

Chapter 3

1. Horgan, 2012, p. 21.

2. Ibid., p. 21.

3. Andrew Bennett (2004). "Case study methods: Design, use, and comparative advantages." *Models, numbers, and cases: Methods for studying international relations* 2.1, pp. 19–55, p. 31.

4. "In the space of a year," after forming, "the Provisionals had effectively superseded the Officials as the main Republican driving force in Northern Ireland." Michael Lawrence Rowan Smith (2002). *Fighting for Ireland?: The Military Strategy of the Irish Republican Movement*. Routledge, p. 91.

5. Ibid., p. 1.

6. Tom Whitehead (December 25, 2015). "Real IRA Could Launch One-Off Attacks on Mainland Britain, Ministers Warned." *The Telegraph*.

7. Richard English (2004). *Armed Struggle: The History of the IRA*. Oxford University Press, pp. 30–31.

8. English, 2004, p. 32.

9. For more on the Troubles, refer to Paul Bew and Gordon Gillespie (1999). *Northern Ireland: A Chronology of the Troubles 1968–1999*. Gill & Macmillan Ltd; Marie-Therese Fay et al. (1999). *Northern Ireland's Troubles: The Human Costs*. Pluto Press; David McKittrick and David McVea (2002). *Making Sense of the Troubles: The Story of the Conflict in Northern Ireland*. New Amsterdam Books.

10. "The Northern Ireland Peace Agreement" (Apr. 1998), pp. 3-5. Available at https://peacemaker.un.org/sites/peacemaker.un.org/files/IE%20GB_980410_Northern%20Ireland%20Agreement.pdf.

11. Ibid., p. 3.

12. *The Starry Plough* (Special Edition), INLA Publication. Dublin: 1977.

13. Bloomer, 1993, p. 5.

14. Ibid., p. 33.

15. Ed Moloney (2003). *A Secret History of the IRA*. Norton, p. 112.

16. J. Bowyer Bell, Jr. (1973). "The Escalation of Insurgency: The Provisional Irish Republican Army's Experience, 1969–1971." *The Review of Politics* 35.3, pp. 398–411.

17. J. Ditch (1977). "Direct Rule and Northern Ireland Administration." *Administration* 25, pp. 328–337.

18. Bernadette C. Hayes and Ian McAllister (2001). "Sowing Dragon's Teeth: Public Support for Political Violence and Paramilitarism in Northern Ireland." *Political Studies* 49.5, pp. 901–922.

19. Holland and McDonald, 1994, p. 12.
20. Ibid., p. 33.
21. Ibid., p. 33.
22. Bloomer, 1993, p. 3.
23. Holland and McDonald, 1994, pp. 28–29.
24. Bloomer, 1993, p. 4.
25. This underscores once again the highly strategic nature of organizational ruptures, and how predicting their onset can be difficult. E.g. Michael O'Higgens (Mar. 1987). "The INLA Devours Itself." *Magill*, pp. 16–26; Bloomer, 1988.
26. Holland and McDonald, 1994, p. 38.
27. Independent Monitoring Commission (2004), Report #1, pp. 12.
28. Bloomer, 1993.
29. Holland and McDonald, 1994.
30. Brian Hanley (Oct. 2010). *The IRA: A Documentary History 1916–2005*. Gill & MacMillan Limited, p. 286.
31. Sophie A. Whiting (May 2016). "Continuity or Dissidence?: The Origins of Dissident Republicans and Their Mandate." *Spoiling the Peace?* Manchester University Press, p. 115.
32. Irish Republican Socialist Committees of North America. "Aims, Principles, and Policies." Republican Socialist Publications. http://www.irsm.org/history/costello/seamus03.html
33. Bloomer, 1993, p. 8.
34. Hanley, 2010, p. 286.
35. English, 2004, p. 177.
36. Holland and McDonald, 1994, p. 47.
37. Ibid., pp. 39–40.
38. Ibid., p. 60.
39. Ibid., p. 63.
40. Bloomer, 1993, p. 8.
41. Ibid.
42. Kiely, Niall (1977). "A man who aroused strong passion." *Irish Times*.
43. O'Higgins, Michael (March 1987). "The INLA Devours Itself." *Magill* 10.7, pp. 16–26.
44. O'Higgins, Michael (March 1987). "The INLA Devours Itself." *Magill* 10.7, pp. 16–26.
45. "Proscription and Deproscription Associations and Organisations/Political Activity." May 1979—December 1980. Northern Ireland Office, Records and Information Management.
46. David Beresrord (March 25,1982). "Feuds breed informers, says Ulster police." *The Guardian*.
47. Holland and McDonald, 1994, pp. 52–53.
48. Holland and McDonnell, 1994, p. 43–44.
49. Ibid., p. 55.

50. Quotation from the IRSP publication, *The Starry Plough*. Quoted in: Bloomer, 1993, p. 2.

51. Though there is some debate as to whether this was an organizational split or really more a form of organizational evolution. Robert W. White (1997). "The Irish Republican Army: An Assessment of Sectarianism." *Terrorism and Political Violence* 9.1, pp. 20–55.

52. John Mooney and Michael O'Toole (2003). *Black Operations: The Secret War against the Real IRA*. Virago Press, p. 38.

53. Tim Pat Coogan (2002). *The IRA*. Palgrave Macmillan, p. 604.

54. English, 2004, pp. 307–308.

55. Quoted in Brendan O'Brien (1999). *The Long War: The IRA and Sinn Féin*. Syracuse University Press, pp. 229–230.

56. English, 2004, p. 310.

57. Ibid., p. 312.

58. Ibid., pp. 313–314.

59. Quoted in ibid., p. 314.

60. *The Mitchell Principles* (May 1996). *The Irish Times*. Available at https://www.irish times.com/news/the-mitchell-principles-1.50976.

61. Mooney and O'Toole, 2003, p. 23.

62. Martyn Frampton (2011). *Legion of the Rearguard: Dissident Irish Republicanism*. Irish Academic Press, p. 91.

63. Mooney and O'Toole, 2003, p. 29.

64. English, 2004, p. 316.

65. Breen, Suzanne (February 3, 2007). "War Back On—Real IRA." *Sunday Tribune*.

66. Okado-Gough, Damien (January 28, 2003). "Interview with the Army Council of the Real Irish Republican Army." Channel 9 TV News, Derry.

67. Mooney and O'Toole, 2003, p. 28.

68. Frampton, 2011, p. 94.

69. James Dingley (2001). "The Bombing of Omagh, 15 August 1998: The Bombers, Their Tactics, Strategy, and Purpose behind the Incident." *Studies in Conflict and Terrorism* 24.6, pp. 451–465, p. 454.

70. Also known as the Good Friday Accords. Independent Monitoring Commission, Report #1, p. 15.

71. Suzanne Breen (February 3, 2007.). "War Back On—Real IRA." *Sunday Tribune*.

72. Mooney and O'Toole, 2003, p. 23.

73. John Horgan and John F. Morrison (2011). "Here to Stay? The Rising Threat of Violent Dissident Republicanism in Northern Ireland." *Terrorism and Political Violence* 23.4, pp. 642–669.

74. Frampton, 2011, pp. 97–98.

75. Ibid., p. 99.

76. Ibid., p. 98.

77. Ibid., p. 100.

78. Mooney and O'Toole, 2003, p. 29.

79. Ibid., p. 29.

80. Frampton, 2011, p. 117.

81. Lisa Langdon, Alexander J. Sarapu, and Matthew Wells (2004). "Targeting the Leadership of Terrorist and Insurgent Movements: Historical Lessons for Contemporary Policy Makers." *Journal of Public and International Affairs* 15, pp. 59–78, p. 68.

82. Ibid., p. 71.

83. Boyne, Sean, (2009). "Fresh Troubles—Dissidents Rise Again in Northern Ireland," in Jane's Intelligence Review.

84. Independent Monitoring Commission (2004), Report #1, p. 15.

85. Shapiro, 2013.

86. John Mooney and Michael O'Toole (2003). Black Operations: The Secret War Against the Real IRA. Virago Press, p. 36.

87. Ibid, p. 37.

88. Mooney and O'Toole, 2003, p. 37.

89. Evan Perkoski (2019). Internal politics and the fragmentation of armed groups. *International Studies Quarterly*, 63.4, pp. 876–889.

90. Price, 2012; Jordan, 2009.

91. Holland and McDonald, 1994.

92. Dingley, 2001.

93. Horgan, 2012.

94. And when the dominant organization shifted from the OIRA to the PIRA, the PIRA initially gave the INLA support. This eventually changed with the PIRA calling for them to disband. Joanne Wright (1990). "PIRA Propaganda: The Construction of Legitimacy." *Journal of Conflict Studies* 10.3, pp. 24–41.

95. J. Bowyer Bell (1979). *The Secret Army: The IRA, 1916–1979*. Academy Press Dublin, p. 421.

96. John McGarry and Brendan O'Leary (2004). *The Northern Ireland Conflict: Consociational Engagements*. Oxford University Press, p. 282.

97. Richards, 2001, p. 80.

98. Andrew Silke (1999). "Rebel's Dilemma: The Changing Relationship between the IRA, Sinn Féin and Paramilitary Vigilantism in Northern Ireland." *Terrorism and Political Violence* 11.1, pp. 55–93, p. 59.

99. Silke, 1999, p. 59.

100. Ross Frenett and M. L. R. Smith (2012). "IRA 2.0: Continuing the Long War—Analyzing the Factors behind Anti-GFA Violence." *Terrorism and Political Violence* 24.3, pp. 375–395, p. 389.

101. Ibid., p. 389.

102. Bloom, 2005.

103. *Parliamentary Debates, Official Report* (June 2005). Volume 605. Dáil Éireann.

104. "Northern Ireland: The Real IRA's Capabilities" (Nov. 2006). Stratfor.

105. James Sturcke (Mar. 2009). "Explainer: Real IRA and Continuity IRA." *The Guardian*.

106. Hayes and McAllister, 2001, pp. 913–914.

107. Peter R. Neumann (2005). "From Revolution to Devolution: Is the IRA Still a Threat to Peace in Northern Ireland?." *Journal of Contemporary European Studies* 13.1, pp. 79–92.

108. Jonathan Tonge (2012). "'No-One Likes Us; We Don't Care': 'Dissident' Irish Republicans and Mandates." *The Political Quarterly* 83.2, pp. 219–226, p. 226.

109. Jonathan Tonge (2004). "'They Haven't Gone Away, You Know'. Irish Republican 'Dissidents' and 'Armed Struggle'." *Terrorism and Political Violence* 16.3, pp. 671–693.

110. Ibid., p. 675.

111. Martyn Frampton (2012). "Dissident Irish Republican Violence: A Resurgent Threat?." *The Political Quarterly* 83.2, pp. 227–237.

112. Maykel Verkuyten (2005). "Ethnic Group Identification and Group Evaluation among Minority and Majority Groups: Testing the Multiculturalism Hypothesis." *Journal of Personality and Social Psychology* 88.1, pp. 121–138; McLauchlin and Pearlman, 2012.

Chapter 4

1. Barbara Geddes (1990). "How the Cases You Choose Affect the Answers You Get: Selection Bias in Comparative Politics." *Political Analysis* 2, pp. 131–150.

2. "Jemaah Islamiyah in South East Asia: Damaged But Still Dangerous" (2003). International Crisis Group Asia Report #63, p. 6.

3. Ari Weil (2017). "The Red Army Faction: Understanding a Measured Government Response to an Adaptive Terrorist Threat." *Cornell International Affairs Review 10.2.* pp. 129–159.

4. Clark, 1984.

5. Bueno de Mesquita, 2008.

6. Ibid., p. 400.

7. Shapiro, 2013.

8. "IS 'caliphate' Defeated but Jihadist Group Remains a Threat" (March 23, 2019). *BBC News.*

9. Thomas Gibbons-Neff (October 8, 2019). "Ally of Al Qaeda Killed in Afghanistan Raid, Officials Say, but Taliban Deny It." *New York Times.*

10. Bilal Y. Saab and Magnus Ranstorm (November 28, 2007). "Fatah Al Islam: How an Ambitious Jihadist Project Went Awry." Brookings.

11. Weil, 2017.

12. S. L. A. Marshall (1947). *Men Against Fire.* New York: William Morrow.

13. Gates, 2002; Eli Berman and David D. Laitin (2008). "Religion, Terrorism and Public Goods: Testing the Club Model." *Journal of Public Economics* 92.10, pp. 1942–1967; Thomas Hegghammer (2013). "The Recruiter's Dilemma: Signalling and Rebel Recruitment Tactics." *Journal of Peace Research* 50.1, pp. 3–16; Daniel Byman (2013). "Fighting Salafi-Jihadist Insurgencies: How Much Does Religion Really Matter?." *Studies in Conflict & Terrorism* 36.5, pp. 353–371.

14. Ervand Abrahamian (1982). *Iran between Two Revolutions*. Princeton University Press, p. 493.

15. Mandel, David R. (2009). "Radicalization: What does it mean," in *Home-grown Terrorism*, eds. Thomas M. Pick, Anne Speckhard, and Beatrice Jacuch. IOS Press, pp. 101–113. John Horgan (2008). "From Profiles to Pathways and Roots to Routes: Perspectives from Psychology on Radicalization into Terrorism." In *The Annals of the American Academy of Political and Social Science* 618.1, pp. 80–94; Thomas Hegghammer (2006). "Terrorist Recruitment and Radicalization in Saudi Arabia." In *Middle East Policy* 13.4, p. 39; Scott Atran (2006). "The Moral Logic and Growth of Suicide Terrorism." In *Washington Quarterly* 29.2, pp. 127–147; John L. Esposito (2002). *Unholy War: Terror in the Name of Islam*. Oxford University Press.

16. Ibid.

17. This is much more appropriate than ordinary least squares, though I present the OLS results in the Appendix. See: Jerald F. Lawless (1987). "Negative Binomial and Mixed Poisson Regression." *The Canadian Journal of Statistics/La Revue Canadienne de Statistique*, pp. 209–225; Paul D. Allison and Richard P. Waterman (2002). "Fixed-Effects Negative Binomial Regression Models." *Sociological Methodology* 32.1, pp. 247–265; Joseph M. Hilbe (2011). *Negative Binomial Regression*. Cambridge University Press.

18. Kydd and Walter, 2006, p. 51.

19. Bloom, 2005, p. 100.

20. Horowitz, 2010; Blair et al., 2022.

21. Asal and Rethemeyer, 2008.

22. J. Lyall (2009). "Does Indiscriminate Violence Incite Insurgent Attacks?: Evidence from Chechnya." In *Journal of Conflict Resolution* 53.3, pp. 331–362; Lyall and Wilson, 2009; Laura Dugan and Erica Chenoweth (2012). "Moving beyond Deterrence: The Effectiveness of Raising the Expected Utility of Abstaining from Terrorism in Israel." In *American Sociological Review* 77.4, pp. 597–624; Evan Perkoski and Erica Chenoweth (2010). "The Effectiveness of Counterterrorism in Spain: A New Approach." In *International Studies Association Annual Meeting, New Orleans, LA*, March, pp. 15–17.

23. Ibid.

24. Chenoweth, 2010.

25. Quan Li (2005). "Does Democracy Promote or Reduce Transnational Terrorist Incidents?." *Journal of Conflict Resolution* 49.2, pp. 278–297; Crenshaw, 1981.

26. Pape, 2003; Pape, 2005.

27. Fearon and Laitin, 2003.

28. Monty G. Marshall, Keith Jaggers, and Ted Robert Gurr. (2011). "Polity IV project: Political regime characteristics and transitions, 1800–2011." *Center for systemic peace*.

29. Kydd and Walter, 2006.

30. There are, however, conditions under which this metric does not accurately capture what it is intended to. There are times when an organization is no longer launching violent attacks but still exists in another capacity. Some militant organizations have transitioned away from violent activities but continue to exist as a political party

or another type of nonviolent organization. An organization could also be on a ceasefire and biding its time before relaunching its armed campaign. The Taliban, for instance, has at various times been a violent nonstate actor and also the government of Afghanistan. These occurrences are relatively rare, so my conceptualization should provide an accurate picture of militant groups' survival. See: Young and Dugan, 2014.

31. Paul D. Allison (2010). *Survival Analysis Using SAS: A Practical Guide*. Sas Institute.

32. S. Brock Blomberg, Khusrav Gaibulloev, and Todd Sandler (2011). "Terrorist Group Survival: Ideology, Tactics, and Base of Operations." *Public Choice* 149.3–4, pp. 441–463.

33. Richard Sosis and Candace Alcorta (2008). "Militants and Martyrs: Evolutionary Perspectives on Religion and Terrorism." *Natural Security: A Darwinian Approach to a Dangerous World*, pp. 105–124.

34. Chenoweth, 2010; Li, 2005; William Lee Eubank and Leonard Weinberg (1994). "Does Democracy Encourage Terrorism?." *Terrorism and Political Violence* 6.4, pp. 417–435.

35. Jason Lyall (2010). "Do Democracies Make Inferior Counterinsurgents? Reassessing Democracy's Impact on War Outcomes and Duration." *International Organization* 64.01, pp. 167–192; Dan Reiter and Allan C. Stam (1998). "Democracy and Battlefield Military Effectiveness." *Journal of Conflict Resolution* 42.3, pp. 259–277; Dan Reiter and Allan C. Stam (2002). *Democracies at War*. Princeton University Press.

36. Young and Dugan, 2014.

37. Marshall, Jaggers, and Gurr, 2013.

38. Robert C. Feenstra, Robert Inklaar, and Marcel P. Timmer (2015). "The Next Generation of the Penn World Table." *American Economic Review* 105.10, pp. 3150–3182.

39. James A. Piazza (May 2011). "Poverty, Minority Economic Discrimination, and Domestic Terrorism." *Journal of Peace Research* 48.3, pp. 339–353; Andreas Freytag et al. (2011). "The Origins of Terrorism: Cross-Country Estimates of Socio-Economic Determinants of Terrorism." *European Journal of Political Economy* 27, S5–S16; Bueno de Mesquita, 2005a.

40. Bloom, 2005, p. 95.

41. Young and Dugan, 2014.

42. Ibid., p. 3.

43. Ibid.

Chapter 5

1. "IS 'caliphate' Defeated but Jihadist Group Remains a Threat" (2019 March 23, 2019). BBC News.

2. Lahoud and Collins, 2016, p. 200.

3. Colin P. Clarke and Assaf Moghadam (Jan. 2018). "Mapping Today's Jihadi Landscape and Threat." In *Orbis* 62.3, pp. 347–371.

4. Weiss and Hassan, 2016, p. 2.

5. Joby Warrick (2015). *Black Flags: The Rise of ISIS*. Anchor, p. 51.

6. Ibid., p. 53.

7. Jeffrey Gettleman (June 9, 2006). "Abu Musab Al-Zarqawi Lived a Brief, Shadowy Life Replete with Contradictions." *The New York Times*.

8. Michael Weiss and Hassan Hassan (2016). *ISIS: Inside the Army of Terror*. Simon and Schuster, p. 10.

9. Aaron Y. Zelin (2014). "The War between ISIS and Al-Qaeda for Supremacy of the Global Jihadist Movement." *The Washington Institute for Near East Policy* 20.1, pp. 1–11, p. 2.

10. Ibid., p. 2.

11. Fishman, Brian (2016a). "Revising the history of al-Qa'ida's original meeting with Abu Musab al-Zarqawi." *CTC Sentinel* 9.10, pp. 28–33, p. 29.

12. Peter Krause (2008). "The Last Good Chance: A Reassessment of US Operations at Tora Bora." In *Security Studies* 17.4, pp. 644–684.

13. Hashim, 2014, p. 70.

14. Brian Fishman (2016b). *The Master Plan: ISIS, Al Qaeda, and the Jihadi Strategy for Final Victory*. Yale University Press, p. 40.

15. Kenneth Katzman (2008). *Al Qaeda in Iraq: Assessment and Outside Links*. Congressional Research Service.

16. Zelin, 2014, p. 6.

17. Zachary Laub and Jonathan Masters (2016). *The Islamic State*. Council on Foreign Relations.

18. Hashim, 2014.

19. Mendelsohn, 2015; Mendelsohn, 2011.

20. Nelly Lahoud, Daniel Milton, and Bryan Price (2014). *The Group That Calls Itself a State: Understanding the Evolution and Challenges of the Islamic State*. Combating Terrorism Center: Military Academy at West Point, p. 12.

21. Bloom, 2004; Bloom, 2005; Kydd and Walter, 2006; Michael C. Horowitz, Evan Perkoski, and Philip BK Potter (2018). "Tactical Diversity in Militant Violence." In *International Organization* 72.1, pp. 139–171.

22. Zelin, 2014, p. 2.

23. Fishman, 2016b, p. 60.

24. Shapiro, 2013.

25. Zelin, 2014, p. 2.

26. Greg Miller (Aug. 2004). "Al Qaeda Finances Down, Panel Says." *LA Times*. https://www.latimes.com/archives/la-xpm-2004-aug-22-na-commish22-story.html.

27. Ibid., p. 2.

28. Jeffrey Pool (Dec. 2004). *Zarqawi's Pledge of Allegiance to Al-Qaeda*. Jamestown Foundation.

29. Fishman, 2016b, p. 59.

30. Fawaz A. Gerges (2017). *ISIS: A History*. Princeton University Press, p. 75.

31. Lahoud, Milton, and Price, 2014, p. 13.

32. Fishman, 2016b, p. 60.

33. Leah Farrall (2011). "How Al Qaeda Works—What the Organization's Subsidiaries Say about Its Strength." *Foreign Affairs* 90.2, p. 128.

34. Danielle F. Jung et al. (2014). *Managing a Transnational Insurgency: The Islamic State of Iraq's "Paper Trail," 2005–2010*. Military Academy at West Point, Combating Terrorism Center, p. 6.

35. Ibid., p. 7.

36. Here, thousands of additional US troops were sent to Iraq with a new mission: to protect and work with the Iraqi population and not, as they had previously done, to solely target insurgent networks and fighters. See: James A. Baker III, Lee H. Hamilton, and Iraq Study Group (2006). *The Iraq Study Group Report*. Vintage.

37. Patrick B. Johnston et al. (2016). *Foundations of the Islamic State: Management, Money, and Terror in Iraq, 2005–2010*. Rand Corporation, p. 18.

38. Mark Wilbanks and Efraim Karsh (2010). "How the 'Sons of Iraq' Stabilized Iraq." *Middle East Quarterly* Fall, pp. 57–70; Andrew Phillips (2010). "How Al-Qaeda Lost Iraq." *Terrorism, Security and the Power of Informal Networks*. Edward Elgar Publishing.

39. Notably, this type of strategy requires significantly more troops to provide continuous security and build relationships with local populations. See: Johnston et al., 2016, p. 23.

40. Ibid., p. 25.

41. Daveed Gartenstein-Ross and Nathaniel Barr (2017). "How Al-Qaeda Survived the Islamic State Challenge." *Current Trends in Islamist Ideology* 21, pp. 50–69, p. 4.

42. Jung et al., 2014; Johnston et al., 2016, pp. 6, 25.

43. Lahoud, Milton, and Price, 2014, p. 13.

44. Interestingly, many scholars argue that the transition of power was surprisingly useful for AQI: the leaders of armed groups are almost constantly under threat, and AQI's long-term viability was improved by its devising legalistic procedures that governed the transfer of power. Not only did this succession from Zarqawi to Al-Muhajir prevent dissent and internal challenges, it also may have been a useful signal to outsiders that the group takes its norms and procedures seriously. This commitment to an orderly succession distinguished AQI from other jihadist organizations, which was especially beneficial at a time when AQI's reputation was already damaged. See: Haroro J. Ingram, Craig Whiteside, and Charlie Winter (2020). *The ISIS Reader: Milestone Texts of the Islamic State Movement*. Oxford University Press, pp. 16–17.

45. Lahoud, Milton, and Price, 2014, pp. 14–15.

46. Ayman Al-Zawahiri (May 2014). *Testimonial to Preserve the Blood of Mujahideen in As-Sham*.

47. Ingram J. Haroro and Craig Whiteside (2019). "Caliph Abu Unknown: Succession and Legitimacy in the Islamic State." *War on the Rocks*.

48. Hashim, 2014, p. 72.

49. Craig Whiteside (2016). "The Islamic State and the Return of Revolutionary Warfare." In *Small Wars & Insurgencies* 27.5, pp. 743–776.

50. Fishman, 2016b, p. 154.

51. Ibid., pp. 116–117.

52. Ibid., pp. 117–118.

53. David Alexander (June 4, 2010). "Qaeda in Iraq Struggling after Leadership Blow—US." *Reuters.*

54. Katzman, 2008, p. ii.

55. Fishman, 2016b, p. 148.

56. Harith Hasan Al-Qarawee (2014). *Iraq's Sectarian Crisis: A Legacy of Exclusion.* Vol. 24. Carnegie Endowment for International Peace. Dylan O'Driscoll (2017). "Autonomy Impaired: Centralisation, Authoritarianism and the Failing Iraqi State." In *Ethnopolitics* 16.4, pp. 315–332.

57. Charlotte F. Blatt (2017). "Operational Success, Strategic Failure: Assessing the 2007 Iraq Troop Surge." *The US Army War College Quarterly: Parameters* 47.1, p. 6; Nick Schifrin (2018). "Campaign Analysis: The 'Surge' in Iraq, 2007–2008." *Orbis* 62.4, pp. 617–631; Samuel Helfont (2018). "An Arab Option for Iraq." *Orbis* 62.3, pp. 409–421.

58. Weiss and Hassan, 2016, p. 91.

59. Tim Arango and Eric Schmitt (August 10, 2014). "U.S. Actions in Iraq Fueled Rise of a Rebel." *The New York Times.*

60. Liz Sly (April 4, 2015). "The Hidden Hand behind the Islamic State Militants? Saddam Hussein's." *Washington Post.*

61. Hashim, 2014, p. 73.

62. As I describe in previous chapters, there were good reasons for Al Qaeda to pursue this strategy: it is less risky than trying to establish its own operations in a new country, and it is much cheaper as well. Interestingly, neither AQ nor ISI opted to embrace their relationship with JN in a meaningful way. Neither group made its sponsorship public, and AQ refrained from adding "Al-Qaeda" to the name of JN as it had done many times before, including with AQI. Instead, Al Qaeda and ISI opted for ambiguity in their Syrian operations. This strategy proved useful to JN during the group's formative year as it was spared from the international attention that the Al Qaeda brand would inevitably bring. See: Mendelsohn, 2015.

63. Regarding the organization's name (AQI versus ISI), Hamming writes that "In the years between 2006 and 2014, it was generally assumed that despite its name change, the Islamic State of Iraq was still Al-Qa'ida's affiliate in Iraq. Even Al-Qa'ida saw it that way, despite being ambiguous in its official communication on the group in Iraq. However, it seems plausible that the Islamic State, when it branded itself by that name in 2006, from that point on considered itself to be independent from Al-Qa'ida as it has been consistently arguing since the split in 2014." Tore Hamming (2019a). *The Hardline Stream of Global Jihad: Revisiting the Ideological Origin of the Islamic State*, p. 6.

64. The group's name has also been translated from Arabic into English as, more precisely, the Islamic State of Iraq and ash-Sham, with ash-Sham referring to Syria or the Levant. For more on this, see Patrick J. Lyons and Mona El-Naggar (June 18, 2014). "What to Call Iraq Fighters? Experts Vary on S's and L's." *The New York Times.*

65. "Translation of Al-Qaeda Statement on Feb. 3, 2014 Acknowledging ISIS Officially Isn't Part of AQ" (Feb. 2014). Jihadology.

66. This sort of conceptualization has analogies in other fields; for instance, when studying the onset of mass killings, researchers have distinguished between "accelerators" and "triggers," with the latter immediately preceding the event. Barbara Harff and Ted Robert Gurr (1998). "Systematic Early Warning of Humanitarian Emergencies." *Journal of Peace Research* 35.5, pp. 551–579.

67. Colin P. Clarke (2019). *After the Caliphate: The Islamic State & the Future Terrorist Diaspora*. John Wiley & Sons.

68. Nicolas Hénin (Sept. 2015). *Jihad Academy: The Rise of Islamic State*. Bloomsbury Publishing.

69. Hamming, 2019a, p. 6.

70. Ibid., p. 3.

71. Brian Fishman (2009). *Dysfunction and Decline: Lessons Learned from Inside Al-Qa'ida in Iraq*. Military Academy at West Point, Combating Terrorism Center, p. 10.

72. Fishman, 2016b.

73. Zelin, 2014, p. 3.

74. "Letter Exposes New Leader in Al-Qa'ida High Command" (Sept. 2006). West Point, NY: Combating Terrorism Center.

75. Ibid.

76. Clint Watts (2016). "Deciphering Competition between Al-Qa'ida and the Islamic State." *CTC Sentinel* 9.7, pp. 1–6, p. 4.

77. "February 2004 Coalition Provisional Authority English Translation of Terrorist Musab al Zarqawi Letter Obtained by United States Government in Iraq" (Feb. 2004). US Department of State.

78. Ingram, Whiteside, and Winter, 2020, p. 13.

79. Lahoud and Collins, 2016, p. 31.

80. Ayman al-Zawahiri (2010). "Zawahiri's Letter to Zarqawi." Combating Terrorism Center at West Point.

81. al-Zawahiri, 2010.

82. Tricia Bacon and Elizabeth Grimm Arsenault (2019). "Al Qaeda and the Islamic State's Break: Strategic Strife or Lackluster Leadership?." *Studies in Conflict & Terrorism* 42.3, pp. 229–263, p. 238.

83. "Letter to Abu Hamzah Asking to Fix Mistakes in the Field" (n.d.). West Point, Combating Terrorism Center.

84. "Letter Exposes New Leader in Al-Qa'ida High Command," 2006.

85. Bacon and Arsenault, 2019, p. 237.

86. Atassi Basma (June 9, 2013). "Qaeda Chief Annuls Syrian-Iraqi Jihad Merger." Al Jazeera.

87. *Dabiq* 1, p. 33. Published by Jihadology. https://jihadology.net

88. *Dabiq* 2, p. 25. Published by Jihadology. https://jihadology.net

89. Harleen K. Gambhir (2014). "Dabiq: The Strategic Messaging of the Islamic State." *Institute for the Study of War* 15.4, p. 9.

90. *Dabiq* 6, p. 53. Published by Jihadology. https://jihadology.net

91. *Dabiq* 6, p. 53. Published by Jihadology. https://jihadology.net

92. *Dabiq* 6, p. 54. Published by Jihadology. https://jihadology.net

93. Brian M. Perkins (2018). "Clashes Between Islamic State and AQAP Emblematic of Broader Competition." Terrorism Monitor 16.18, The Jamestown Foundation.

94. Haroro J. Ingram (2016). "An Analysis of Islamic State's Dabiq Magazine." In *Australian Journal of Political Science* 51.3, pp. 458–477.

95. Gambhir, 2014, pp. 2, 6.

96. Bloom, 2005.

97. Lorne L. Dawson and Amarnath Amarasingam (2017). "Talking to Foreign Fighters: Insights into the Motivations for Hijrah to Syria and Iraq." *Studies in Conflict & Terrorism* 40.3, pp. 191–210; Gambhir, 2014.

98. Brendan I. Koerner (Apr. 2016). "Why ISIS Is Winning the Social Media War." Wired.

99. Anonymous (August 13, 2015). "The Mystery of ISIS." *In The New York Review.*

100. Hamming, 2019a, p. 5.

101. Joseph Holliday (March 8, 2013). "Syria Update 13-01: Iraq-Syria Overland Supply Routes." Institute for the Study of War.

102. Shane Harris (August 21, 2014). "The Re-Baathification of Iraq." Foreign Policy.

103. Sly, April 4, 2015.

104. Shapiro, 2013.

105. Sudarsan Raghavan (April 14, 2019). "With the ISIS Caliphate Defeated in Syria, an Islamist Militant Rivalry Takes Root in Yemen." *Washington Post.*

106. Daniel Byman (May 2016). "Should We Be Worried about a Merger between ISIS and Al-Qaida?" Slate.

107. Asaf Day (Dec. 11, 2018). "ISIS and Al-Qaeda Clash Not Only on the Battlefield, but Also in Their Propaganda." *The Defense Post.*

108. Daniel Paquette and Joby Warrick (Febraury 23, 2020). "Isis and Al-Qaeda Join Forces in West Africa." *The Independent.*

Chapter 6

1. Ethan Bueno de Mesquita (2007). "Politics and the Suboptimal Provision of Counterterror." *International Organization* 61.01, pp. 9–36.

2. Kydd and Walter, 2006; Greenhill and Major, 2007; Stedman, 1997.

3. James Dobbins, Jason H. Campbell, Sean Mann, and Laurel E. Miller. *Consequences of a Precipitous US Withdrawal from Afghanistan.* Rand Corporation, 2019.

4. Hudson, 1999.

5. Cunningham, 2011.

6. Dugan and Chenoweth, 2012; Perkoski and Chenoweth, 2010; Gary LaFree, Laura Dugan, and Raven Korte (2009). "The Impact of British Counterterrorist Strategies on Political Violence in Northern Ireland: Comparing Deterrence and Backlash Models." *Criminology* 47.1, pp. 17–45.

7. Hudson, 1999.

8. Alec Worsnop (2017). "Who Can Keep the Peace? Insurgent Organizational Control of Collective Violence." *Security Studies* 26.3, pp. 482–516; Paul Staniland (2014). *Networks of Rebellion: Explaining Insurgent Cohesion and Collapse.* Cornell University Press; Weinstein, 2005; Donatella Della Porta (1995). "Left-Wing Terrorism in Italy." *Terrorism in Context*, pp. 105–159; Corinne Bara (2018). "Legacies of Violence: Conflict-Specific Capital and the Postconflict Diffusion of Civil War." *Journal of Conflict Resolution* 62.9, pp. 1991–2016; Steven A. Zyck (2009). "Former Combatant Reintegration and Fragmentation in Contemporary Afghanistan: Analysis." *Conflict, Security & Development* 9.1, pp. 111–131; Kent Layne Oots (1989). "Organizational Perspectives on the Formation and Disintegration of Terrorist Groups." *Studies in Conflict & Terrorism* 12.3, pp. 139–152.

9. Weinberg, Pedahzur, and Perliger, 2008; Anisseh Van Engeland and Rachael M. Rudolph (Apr. 2016). *From Terrorism to Politics.* Routledge; Richards, 2001; Braithwaite and Cunningham, 2020.

10. For example, consider notable work by: Horowitz, 2010; Bader Araj (2008). "Harsh State Repression as a Cause of Suicide Bombing: The Case of the Palestinian–Israeli Conflict." In *Studies in Conflict & Terrorism* 31.4, pp. 284–303; Ami Pedahzur (2005). *Suicide Terrorism.* Cambridge University Press; Bloom, 2004; James A. Piazza (2008). "A Supply-Side View of Suicide Terrorism: A Cross-National Study." In *Journal of Politics* 70.1, pp. 28–39; S. J. Wade and D. Reiter (Apr. 2007). "Does Democracy Matter?: Regime Type and Suicide Terrorism." In *Journal of Conflict Resolution* 51.2, pp. 329–348; Robert T. Holden (1986). "The Contagiousness of Aircraft Hijacking." In *American Journal of Sociology*, pp. 874–904; Philip Potter, Evan Perkoski, and Michael C. Horowitz (2013). "The Life-Cycle of Terrorist Tactics: Learning from the Case of Hijacking." Working Paper; Laura Dugan, Gary LaFree, and Alex R. Piquero (2005). "Testing a Rational Choice Model of Airline Hijackings." In *Intelligence and Security Informatics.* Springer, pp. 340–361.

11. Mapping Militant Organizations (July 2018). *Tehrik-i-Taliban Pakistan.* Stanford University.

12. Mia Bloom (2017). "Constructing Expertise: Terrorist Recruitment and 'Talent Spotting' in the PIRA, Al Qaeda, and ISIS." *Studies in Conflict & Terrorism* 40.7, pp. 603–623; Khuram Iqbal and Sara De Silva (2013). "Terrorist Lifecycles: A Case Study of Tehrik-e-Taliban Pakistan." *Journal of Policing, Intelligence and Counter Terrorism* 8.1, pp. 72–86; Matthew Dixon (2020). "Militants in Retreat: How Terrorists Behave When They Are Losing." *Studies in Conflict & Terrorism*, pp. 1–25; Dipak K. Gupta (2020). *Understanding Terrorism and Political Violence: The Life Cycle of Birth, Growth, Transformation, and Demise.* Routledge.

13. Olivier J. Walther and Patrick Steen Pedersen (Apr. 2020). "Rebel Fragmentation in Syria's Civil War." *Small Wars & Insurgencies* 31.3, pp. 445–474.

14. "Black Power," in *King Encyclopedia.* Stanford: The Martin Luther King, Jr. Research and Education Institute, Stanford University. https://kinginstitute.stanford.edu/encyclopedia/black-power.

15. Mary-Alice Waters (1969). "The Split at the SDS National Convention." In *The Militant* 33.27.

16. Erica Chenoweth and Maria J. Stephan (2011). *Why Civil Resistance Works: The Strategic Logic of Nonviolent Conflict*. Columbia University Press.

17. Paquette and Warrick, 2020.

18. Charles W. Mahoney (2017). "Splinters and Schisms: Rebel Group Fragmentation and the Durability of Insurgencies." *Terrorism and Political Violence*, pp. 1–20.

19. Morrison, 2013; Bueno de Mesquita, 2008.

20. Doctor, 2020.

Index

Page numbers followed by *t*, *f*, and *n*, respectively, indicate tables, figures, and notes.